Studies in Rhythm Engineering

This is a multi-disciplinary book series, ranging from astrophysics to biology, chemistry, mathematics, geophysics and materials science. Its primary scope is the fundamental science and associated engineering wherever cyclic and rhythmic oscillations are observed.

Time neither being an entity nor a process is unmeasurable and undefined, although a clock only measures the passage of time. The clock drove recurring processes are observed in the biological rhythms, astrophysical and geophysical environments. Always, the clocks are nested, arranged in a geometric shape to govern a phenomenon in nature. These clocks are made of atoms, molecules, their circuits, and complex networks. From biology to astrophysics, the clocks have enriched the science and engineering and the series would act as a catalyst and capture the forthcoming revolution of time cycles expected to unfold in the 21st Century. From the cyclic universe to the time crystal, the book series makes a journey through time to explore the path that time follows.

The series publishes monographs and edited volumes.

More information about this series at https://link.springer.com/bookseries/16136

Pushpendra Singh · Kanad Ray ·
Anirban Bandyopadhyay

Biological Antenna to the Humanoid Bot

Electromagnetic Resonances in Biomaterials

Springer

Pushpendra Singh
International Institute of Invincible
Rhythms IIIR
Shimla, India

Kanad Ray
Physics & Electronic and Communication
Amity School of Applied Science
Jaipur, India

Anirban Bandyopadhyay
National Institute for Materials Science
Tsukuba, Japan

ISSN 2524-5546 ISSN 2524-5554 (electronic)
Studies in Rhythm Engineering
ISBN 978-981-16-9676-3 ISBN 978-981-16-9677-0 (eBook)
https://doi.org/10.1007/978-981-16-9677-0

This Springer imprint is published by the registered company Springer Nature Singapore Pte Ltd.
The registered company address is: 152 Beach Road, #21-01/04 Gateway East, Singapore 189721, Singapore

Preface

For the last decade, we have been studying biological systems in a very different manner. Conventionally, it is a trend to measure biological systems as if they are market available electronic and photonic devices. Thousands of literature are published in that direction. However, no one tried to understand and invent the true information structure of a biological system. We started measuring the information in the biological system adopting a Hindu philosophy that fundamental to this universe is eternal vibration, Brahman; which is advocated by western scientists as panpsychism. One premise of this proposal is similar to the "standard genetic code" which states that "one eternal vibration has become everything," which means Brahman has become everything. So, if we want to apply it to unravel the mystery of biological information processing, we need to measure the pure signal generated by the biological system without perturbing them. Then we have to investigate intricately if all the vibrations are connected.

It was a difficult journey because of the fundamental inherent philosophy that we will not pump massive energy to hurt the biosystem and analyze its scream to understand its inner voice. On the contrary, we would understand its language and make a conversation, thus, make an effort to understand its thoughts and mind. The analogy is humane yet helpful to understand our approach to learning the biological system's true properties. The challenge has been our education and training to do research. It teaches us to follow the crowd, follow the existing science culture where lots of instruments are already available, buy them, and measure. Instead, we had to invent atomic probes. Develop wireless characterization protocols. We excite biosystems with ultralow power noise and then measure output using monochromatic polarized light so that "non-demolition" of inner signals is honored with the highest priority. Dielectric resonance microscopy, where we could visualize deep inside a biological system without making any contact. Not just tools, we had to create an infrastructure where we could converse with a biomaterial in one-time domain and listen to it in all other time domains.

The first three chapters have accounted for three fundamental observations that are our original contribution to biophysics. The first chapter suggests that the evolution of the virus teaches us the most fundamental mystery of evolution, a mathematical formulation more generic than Darwin's evolution theory. The second chapter suggests that the noise which bursts in the shape of the geometric structure is the signal that the biological system understands better than ordered signals. Noise bursts need water cover around every single biological system to filter them into meaningful clocks. In the third chapter, biomaterials turn selectively invisible and visible. That is how a simple-looking biomaterial could act as a superposition of many 3D clock assemblies.

Using these three chapters, we have made a journey to build a humanoid robot, where the only engineering that we explore is building clocks or 3D clock assemblies. The sensor makes 3D clock assemblies, integrates them, and processes them with 3D clock assemblies of biological organs, and then decisions are converted. This is a primitive effort to create a civilization of technologies where bits and Turing machines do not rule, but Vedic and Brahmic science-derived technologies reign.

Shimla, India
Jaipur, India
Tsukuba, Japan

Pushpendra Singh
Kanad Ray
Anirban Bandyopadhyay

Contents

About the Authors

Pushpendra Singh is a research executive at the International Institute of Invincible Rhythms. He has done his PhD jointly with Amity University and National Institute for Materials Science (NIMS), Japan, in 2020. He has published 32 papers and currently developing the world's largest brain database "Lotus" (Large operational transhuman understanding system) in NIMS, Tsukuba, Japan. He has built a humanoid bot and a complete time crystal model for the human brain.

Kanad Ray (Senior member, IEEE) received his M.Sc. degree in physics from Calcutta University and a Ph.D. degree in physics from Jadavpur University, West Bengal, India. He has been a Professor of Physics and Electronics and Communication, and is presently working as Head of the Department of Physics, Amity School of Applied Sciences, Amity University Rajasthan (AUR), Jaipur, India. His current research areas of interest include cognition, communication, electromagnetic field theory, antenna and wave propagation, microwave, computational biology, and applied physics. He has been serving as Editor for various Springer book series. He was the Associate Editor of the Journal of Integrative Neuroscience (The Netherlands: IOS Press). He has been Visiting Professor to UTM & UTeM, Malaysia, and Visiting Scientist to NIMS, Japan. He has established MOU with UTeM Malaysia, NIMS Japan, and the University of Montreal, Canada. He has visited several countries such as Netherlands, Turkey, China, Czechoslovakia, Russia, Portugal, Finland, Belgium, South Africa, Japan, Singapore, Thailand, and Malaysia for various academic missions. He has organized various conferences such as SoCPROS, SoCTA, ICOEVCI, and TCCE as General Chair and a steering committee member.

Anirban Bandyopadhyay is a Principal Research Scientist at the National Institute for Materials Science (NIMS), Tsukuba, Japan. He received Ph.D. in Supramolecular Electronics at the Indian Association for the Cultivation of Science (IACS), Kolkata, on 2005. From 2005 to 2008, he was ICYS research fellow at the ICYS, NIMS, Japan, and worked on the brain-like bio-processor. In 2008, he has joined as a permanent scientist at NIMS, working on the time crystal model of the human brain and design-synthesis of brain-like organic jelly, written a book entitled “Nanobrain: The making of an artificial brain from a time crystal,” on 2020. From 2013 to 2014, he was a visiting scientist at the Massachusetts Institute of Technology (MIT), USA. He has received Hitachi Science and Technology award, 2010, Inamori Foundation award, 2011–2012, Kurata Foundation Award, Inamori Foundation Fellow, 2011, and Sewa Society International Member, Japan.

Chapter 1
Is Coronavirus Pushing Humanity Towards an Evolutionary Jump as It Did Long Back When It Initiated Making a Brain?

1.1 Introduction: Seeing Life Using a Language of Clocks

1.1.1 When a Virus Was Attacked by Another Virus: Advantage of Clocks

Thirty thousand years back, when humans were still primitives, the viruses were more life-like; they could regenerate themselves. The surprising part is that the virus is attacked by another virus. Tiny satellite virus Sputnik attacked giant Mamavirus, even a few years back. We could extrapolate the situation to 1.5 million years when an early human came to this planet when the difference between living and nonliving forms of life was minimal. Life form now begins with one embryonic cell, replicating over and over. The process of replication is the evolution of genetic code that also evolves and becomes more complicated. Unfortunately, the evolution of genetic codes does not consider the hierarchy of genetic codes [1]. DNA code holds a history of life form in our planet, and that history maps how we humans have reached the shape we hold now and what we would become.

We have created a new language where a 3D assembly of clocks could rewrite every piece of nature more intricately without losing strong correlations. We believe that just like in the 1940s, the movement of Shannon's information theory taught us to rewrite every single event in nature, rejecting bits and the concept of Turing machines. Our new language enables us to emulate complex geometric shapes as a cage on which nature encodes information. There are plenty of advantages of using finite geometric shapes to recreate all events instead of using bits or a pair of states. A pair of states require human imagination, but a geometric shape could hold astronomical correlations between events. Thus, geometric shapes made of clocks fuse structural symmetry with information content. At the same time, the information structure is an engine that runs to predict the future. To consolidate and predict the future, we have also introduced the concept of representing geometric symmetry using the pattern that primes use to create an integer. We call it phase prime metric,

P. Singh et al., *Biological Antenna to the Humanoid Bot*, Studies in Rhythm Engineering,
https://doi.org/10.1007/978-981-16-9677-0_1

PPM. Our objective is to understand the evolution of life using primes. Here we would make a journey from understanding viruses to humanoid robotics as engineering of primes.

1.1.2 Basic Concepts of the Virus Evolution

The virus is a small parasite made from genetic materials and involved in the proteins and sometimes in the outer membrane. It hijacks the functional mechanism of the living cell to reproduce itself. Every day, we contact the virus, but our immune system prevents us from the majority of the viruses. Maximum from those which we have been vaccinated before. Virus evolution is always an interesting and albeit murky topic for cell biologists and virologists. There is huge diversity among them. Researchers always struggle to classify them and how to relate them to the conventional tree of life. We may consider them as genetic elements and have the ability to move between cells. Initially, viruses are free-living organisms that become parasites. As we know, there are different forms of viruses. They contain other biological organisms such as herpes virus contains DNA genomes, poliovirus contains RNA genomes, and they have different structure and replication strategies. Some viruses have single-stranded genomes like the influenza virus, whereas some have double-stranded genomes like smallpox. Virus shared some features such as:

1. Generally, they are found in tiny sizes and have a small diameter of less than 200 nm.
2. They can produce identical copies to themselves within the host cell.
3. No known virus contains ribosomes, an important component of the cells responsible for making translational machinery.

1.1.3 Do Viruses Have Life

To address this question, we have to be sure about the meaning of life. Although there is no specific definition, biologists always consider that living organisms have some key features: grow, replicate, maintain internal homeostasis, carry out various metabolism processes, and respond to external stimuli. Do viruses hold the same features? We probably realize that they reproduce in the same way. For example, we can be infected by the emitting particles during inhalation when any person coughs. After several days, we become sick, and viruses replicate in our bodies. In other words, we realize that the virus evolves with time. We have to be flu vaccinated every year because the influenza virus evolves/changes from year to year [2]. Viruses do not evolve with the metabolic process and cannot generate the Adenosine triphosphate, ATP. Viruses do not evolve with the translational machinery, do not possess the ribosomes, and cannot independently generate the protein from the RNA molecules,due to these restrictions, viruses replicate within the host cell. In other words, they are intracellular parasites. So from the critical point of view, they are nonliving. Viruses are different types of elements of the life tree, known as capsid-encoding organisms (CEOs).

1.2 Origin of Viruses

There has been a debate among the virologists about the origin of viruses, and there are three main hypotheses.

1.2.1 Progressive Hypothesis

The progressive hypothesis states that viruses arise from the genetic material and can move between the cells. If viruses originated through the progressive hypothesis, they would be the piece of the genetic material capable of walking in the genome and gaining the ability to exit from one cell to another. For example, in retroviruses, which consist of the single-stranded RNA genome, when a virus enters the host cell, a viral enzyme (reverse transcriptase) converts the single-stranded RNA into double-stranded DNA. Such DNA moves to the nucleus of host cells, and another viral enzyme insert the newly formed DNA into the genome of the host cell. RNA polymerase of the host cell generates some new copies of the single-stranded RNA genome of the virus. Progeny viruses assemble and exit from cells to begin the life-like cycle again.

1.2.2 Regressive Hypothesis

The regressive hypothesis states that viruses are the remnants of cellular organisms. According to the regressive hypothesis, the virus may originate from a regressive process. Microbiologists believe that certain bacteria like Rickettsia and Chlamydia species are intracellular parasites and evolved from living ancestors. Some studies suggest that mitochondria of rickettsia species and eukaryotic cells may have a common living ancestor [3], which means viruses may originate from more complex living organisms that lost the information during the replication process.

1.2.3 Virus First Hypotheses

The first hypothesis about the virus is that the virus already evolves with the present cellular hosts. Virus first hypotheses assume that the virus existed before the prokaryote cell. Several studies reported that viruses might be the first replicating entities in a precellular world. Such units became more organized and took the complex's shape over time [4]. Accidently, enzymes evolved to synthesize cavities made of walls like membranes, resulting in the formation of the cells. Viruses may exist before bacteria and archaea [5]. Most biologists believe that the first replicating molecules consist of RNA, not DNA. Some RNA molecules like ribozymes show enzymatic properties, and they behave as catalyzed for the chemical reactions. Somehow, replicating DNA molecules happens before the cell formation and forms the ability to infect the first cell. Researchers are always excited to know the mystery of the origin of life. Currently, there is no clear explanation for the origin of viruses. Understanding the evolutionary history of the virus may help to understand it.

1.3 The Extraordinary Role of Viruses in the Shaping of the Human Evolution

Viruses infect us from the common cold to AIDS and covid-19, but they also have a key role in the evolution of the homo sapiens. We know very well that viruses (Zika, Flu, Ebola, Coronavirus, etc.) affect humanity. How the Zika virus infects the brain in adults is shown in Fig. 1.1. They make us sick, but over millions of years, we have controlled those. From the early stage of life, viruses lead to huge infections in the human species.

1.3.1 Working Mechanism of Virus

A virus is more than a string of genes (made from RNA molecules), and their working mechanism follows the same basic route. When a virus infects the cell, it hijacks its molecular machinery and starts copying its genes. Newly created genes are responsible for collecting new viruses that move out to find new cells for attacking. For many viruses like flu, they are limited. However, retroviruses, HIV, interfere with our DNA (Fig. 1.2). Suppose they are randomly inserted into the organism's genome; they remain there until the virus begins to production again. Once retrovirus gets the organism's DNA, then there is no guarantee that it will stay there. We can read the genetic instructions from the embedded virus, which converts into DNA, and generated copies are pasted at the different places of the genome. Such a process repeats again and again and builds up several copies of viral DNA. These viral DNA get trapped inside the genomes and change randomly. Some move here and there while others get stuck there while leaving. If something happens in the germ cells which create the sperm and eggs, they will pass through the whole generation process and accidentally become part of the organism's genome.

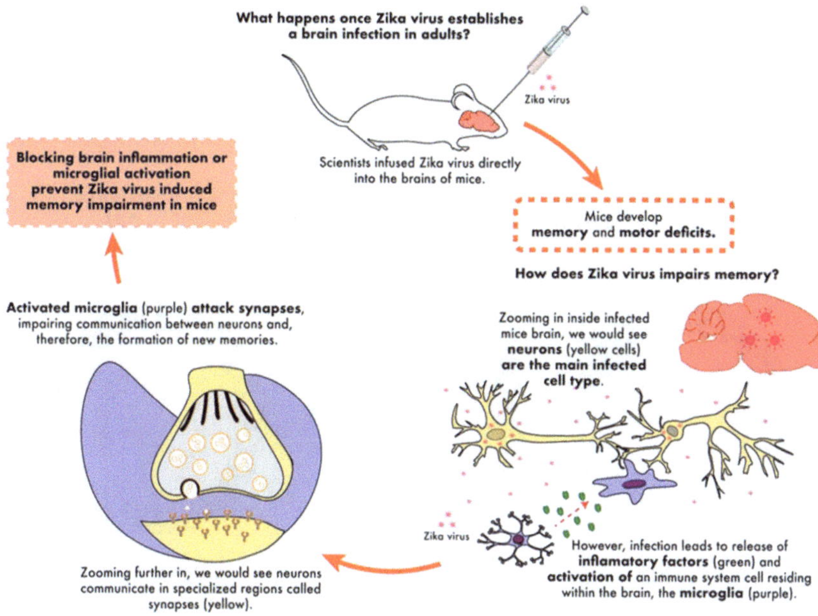

Fig. 1.1 Impact of Zike virus in mice brain. Such a study is carried out on a mice brain. The infection is caused by neurological disorders like memory and cognitive deficit in the mice's brain. https://www.eurekalert.org/multimedia/pub/210245.php

1.3.2 Evolution of Syncytin

Syncytin is the human gene that activates in the placenta. It forms a molecule that fuses the placental cells and creates a layer of tissue, syncitium. Syncytin looks like the gene of the retrovirus. Syncytin viruses are found in humans and other large primates and could not be found in other mammals by similar fusion of cell layers in the placenta. Two syncytin genes are also found in the mice, and their working mechanisms are the same as humans, but they look like different viruses. In some other mammals like cats and dogs, different syncytin genes are there. Clearly, all such mammals were infected by a particular virus. Those viruses play a key role in placental growth and fix permanently in our genome. While horses and pigs do not have a fusion cell layer in their placenta and no viral-derived syncytins, it may be that they never get these fusing viruses.

Viruses may play a key role in the development of the human placenta.

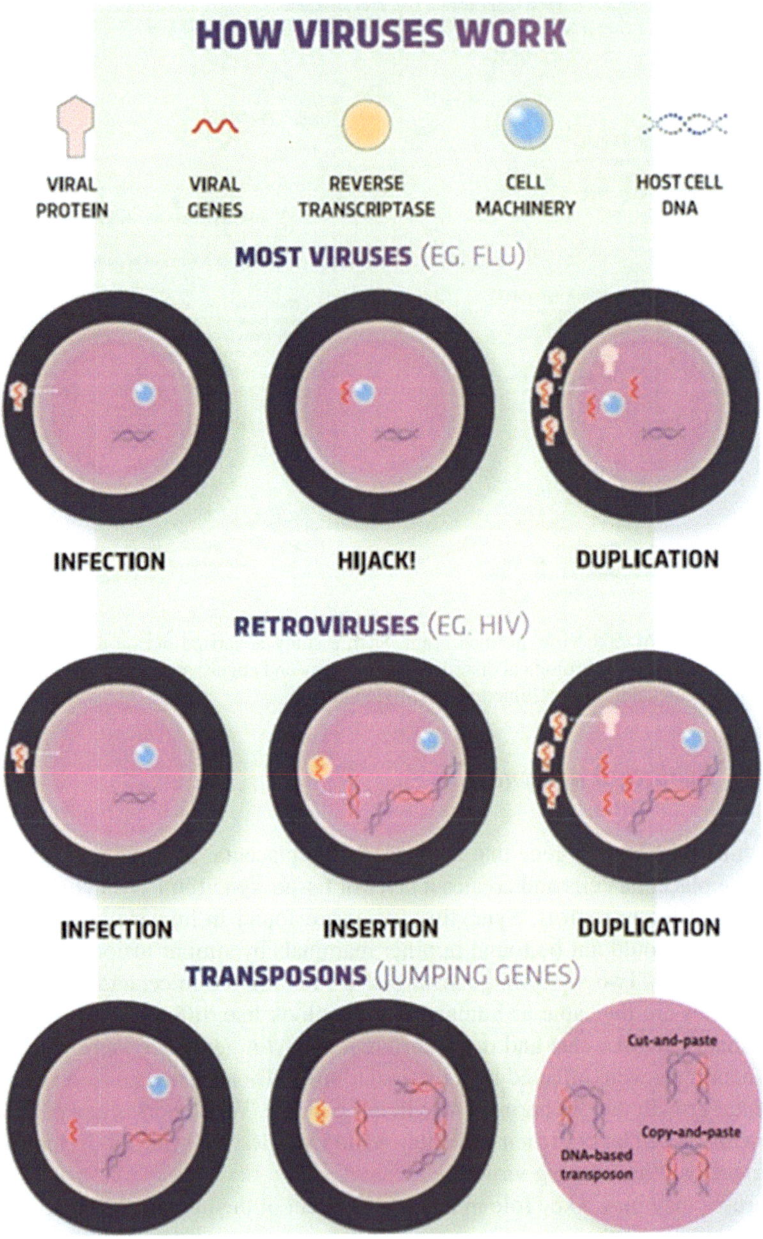

Fig. 1.2 A schematic representation of the working mechanism of the various viruses; I row-EG, FLU; II-EG, HIV, and III-jumping genes. https://www.sciencefocus.com/the-human-body/virus-human-evolution/#:~:text=In%202016%2C%20scientists%20at%20the,which%20is%20the%20'danger%20signal

1.3.3 Humans and Viruses Both Live All Over the Food Pyramid

Humans and viruses both eat almost everything; they attack almost all members of the food pyramid. So in a food pyramid, both viruses and humans live together in all positions of a food pyramid. Therefore, it is not far that both humans and viruses would fight against each other to dominate the food pyramid.

1.3.4 How Do Viruses Jump from Animals to Humans?

Every single animal hosts a particular virus that infects us. Over time, these viruses jump to humans and are known as zoonotic viruses. As human populations grow, they move in wider space. There is a high probability of contact with the animals compared to what they normally have. Viruses jump from animal to human through close contact with their body fluids like blood, muscle, or urine, in the same way as humans do to humans. A unique virus evolves with a particular species, and it rarely happens. When the virus jumps to another species, it needs a large dosage of contact with the virus. In starting, the virus is not friendly to the new host. However, over time, it can produce variants. Viruses are often the cause of many diseases through the zoonosis process because viruses and initial hosts develop together. So species have sufficient time to build up resistance, while on the other hand, new species may not have the sufficient ability to tackle the virus. For example, when we come in touch with the bats and their viruses. We may get rabies or Ebola virus disease, but they are less affected. Bats are the source of several diseases like the SARC-CoV (2003), SARC-CoV-2, MERS-CoV (2012). All these viruses jump from bats to animals by the intermediate animals.

1.3.5 How Does a Virus Make Us Ill?

We get the infection when a virus penetrates the skin's physical barrier or through the mucus, entering a suitable cell. Once inside, it takes over the cells and forces them to replicate themselves, resulting in damage or death to the cell. Newly-made viruses find the new cell for the replication process. When we get ill, the viruses have infected many cells, so the normal functioning of our body changes. Normally, viruses always infect a specific place of our body. Rhinoviruses infect us by upper airways behind the nose, resulting in the cold, while Covid-19 infects the lower airways and lungs. Our body fights against the viruses by collecting strong responses by calling the special cells from tissues and organs. Special cells mean some cells which behave as an antibody against viruses. Some kill the infected cells, and some keep the memory of the virus for the next encounter. Sometimes we feel sick (fever, flu, etc.) which is a result of our body fighting against the virus, not because of the virus.

1.3.6 How Long a Virus Can Live Outside of the Body

Viruses can live outside of the body. How long can they stay? How long they can stay depends on the humidity and temperature of the environment, they can stay longer on water-resistant surfaces like plastic and stainless steel. A cold virus survives for a long time on the indoor surface while its infection ability reduces over time. Flu viruses survive in the air for several hours at a low temperature, and on the hard surface, they can survive for 24 h. Norovirus and hepatitis have a long period of survival (almost one week) on the surface under suitable conditions.

1.4 How Are Viruses Important for Our Brain

Over a long time, retroviruses have been incorporated into the human DNA, making up 10% of the total genome. Retroviruses may have an impact on gene expressions and play a significant role in human brain development as well as neurological diseases. Signature of endogenous retroviruses (ERV) exists in that part of DNA previously considered as junk DNA. Currently, researchers are interested in these parts. The genes that control protein production in our body show a smaller proportion of DNA (2%) than ERV (8–10%) of the total genome. ERV affects protein production, a new source of information about the human brain. Involved retrovirus in our genome may serve as the docking platform for a co-repressor protein TRIM28. This protein can switch off the virus and standard gene associated with the DNA helix and affect gene expression. Different peoples belong to different switching-off mechanisms. Since retroviruses are located at different locations in the genome, it is a key tool for evolution. EVR has a regulatory role in neurons; such studies were based on mice and human cells. Our brain functions are affected by the EVR; the functional mechanism they affect may contribute to human brain development (Fig. 1.3; [6, 7]).

1.4.1 How Virus Can Affect the Human Brain Cell

Viruses can affect the human brain cells when SARC-CoV-2 is introduced into the human mini-brain, where an infection was found. Mini brain means the tiny tissue cultures prepared from human brain cells that simulate the whole organ. 36% of the patients of COVID-19 show neurological symptoms, but it is not clear whether the virus infects the human brain cells or not! Certain human neurons have the same receptor-like SARS-CoV-2 enter to lungs. ACE2 may also provide information

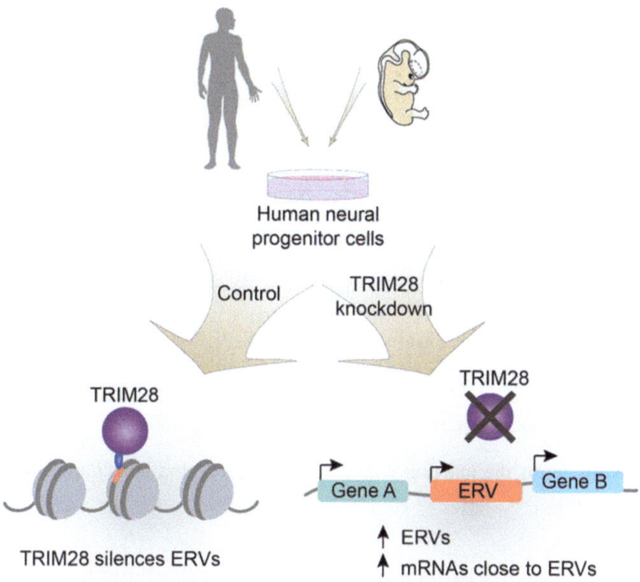

Fig. 1.3 Functional mechanism of EVR during the human brain development [6]

regarding the brain. The human brain is very well shaped; its blood–brain barrier prevents the virus, chemical agents, and bacteria. Although, it was observed that serious swelling in COVID-19 patients shows disintegrated blood barrier. On the other hand, the second concern is how SARS-CoV-2 makes an impact on the developing brain. Some studies show that the virus crosses the placenta, and during the early stage of brain development, embryos lack a blood–brain barrier. However, there is no evidence that the virus leads to brain development disorder. The mini-brain, which is a mimic version of the growing human brain, consists of ACE2 receptors at the early stages of human brain development, so, during the pregnancy, extra precautions should be there. Such study has another important application in understanding the characteristics of symptoms so we can tackle the COVID-19 disease.

1.4.2 Effect of Poliomyelitis Virus on the Human Brain Cells

Poliomyelitis virus I affect the human brain cells grown in the culture plate and may cause cell death in 3 days. Neurons treat viruses differently; some die, and some contact their neurites. At the tip of neurites, there were terminal bulbs when they were pulled in the cell body. The virus affects the oligodendroglia, astrocytes, and macrophages more quickly than the neurons. The junction formation between Poliovirus and neuron's receptor is shown in Fig. 1.4. For the cell in good condition, the cell culture of the individual donor of brain tissue matters, not the age of cell

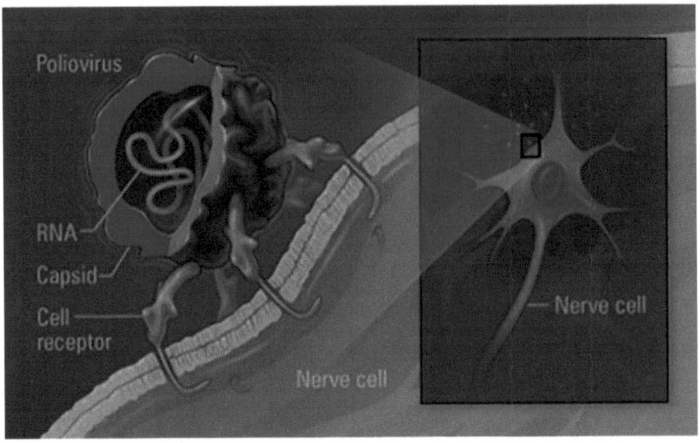

Fig. 1.4 Poliovirus forms the junction with the neuron's receptor. https://amhistory.si.edu/polio/virusvaccine/how.htm

culture. The fetal brain cells were more sensitive to the virus than the adult brain cells. After the inoculation, fetal brain cells reacted within 30 min while the adult neurons consumed 2 24 h. This happens due to the virus infection because they were protected by antiserum [8].

1.5 How Covid-19 Affects the Human Nervous System?

Neurological disorders have emerged because of the ongoing COVID-19 pandemic. Except for respiratory insufficiency, many covid-19 patients suffer from headaches, confusion, stroke problems, disorientation, agitation, and loss of smell. It also affects the nervous system for a long time [9]. The mechanism of neuropathogenesis SARS-CoV-2 is related to the chronic neurological results of the infection [10]. It was also seen that the patient who suffered from covid-19 had swelling in the brain tissue [11]. Other studies describe the myelin deterioration which protects the neuron [12]. Although viruses can infect the brain and spinal cord, it is not clear whether it is due to SARC-CoV-2 or any other. It may be the result of overstimulation of the immune system. Finding out the disease mechanism is important because both cases require different treatment. Some seriously affected people had only respiratory symptoms, and peripheral nerves were also affected by it. SARS (severe acute respiratory syndrome) and MERS (middle east respiratory syndrome) viral diseases are caused by the coronavirus.

SARC-CoV-2 can infect the neurons by killing and reducing the synaptic junction formation between them [13, 14]. Olfactory nerves may be the possible route for viruses to reach people's brains (Fig. 1.5), but the evidence is found against it. Few studies reported that viruses are detected in the brains of covid patients using electron

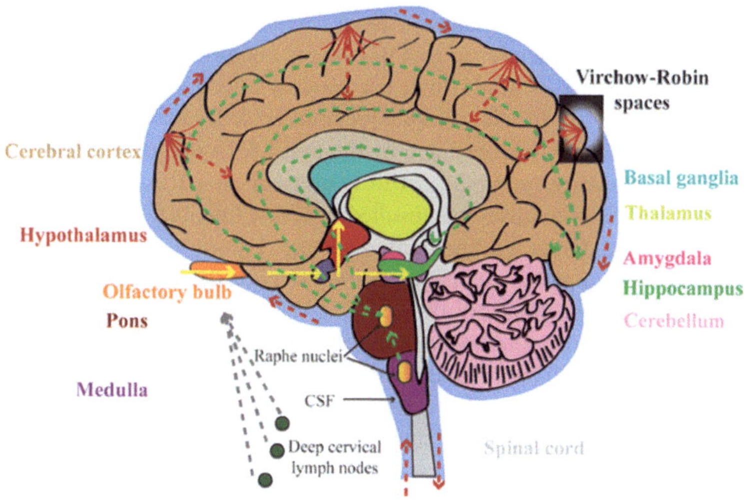

Fig. 1.5 Possible route of hCoVs virus entry in the human brain [16]

microscopes, but they are detected at a very low level and were not found consistent. If viruses attack the olfactory nerve, the associative brain region should be affected first, but it was not found. Small infections are seen in the brain, but clusters form around the blood vessel [13]. Finding the virus in the cerebrospinal fluid and the brain and spinal cord is difficult other than organs even by using polymerase chain reaction, PCR. One reason may be that the ACE2 receptor, a protein on the cell from which viruses enter, is not found much in brain cells [15].

All possible ways of CNS infection with hCOVs in the human brain are shown. Routes with solid arrows were detected experimentally, while the routes with dots were speculative ones. Yellow route: Olfactory bulb to the olfactory cortex to the hippocampus to the amygdala or the hypothalamus. Green route; serotoninergic dorsal raphe system. The red route includes the hematogenous and Virchow-Robin spaces. The Gray route passes through the lymphatic system. Color codes of brain components and texts are the same.

1.5.1 Artificially Triggering Virus Evolution: Wuhan's Lab Was Experimenting with Junction Optimization

Around December 2019, a novel coronavirus (CoV) started to circulate among humans in Wuhan, China. Initially, the virus's effect on humans was unknown. Since then, this virus, named "severe acute respiratory syndrome coronavirus 2" (SARS-CoV-2), has been the cause of a global pandemic, with over 115 million confirmed cases worldwide and there had been 2.56 million fatalities (https://www.worldometers.info/coronavirus/). SARS-global CoV-2's spread and devastating effects have made it one of the most well-known pandemics in human history. Unfortunately, some aspects of the pandemic suggest that this is not a one kind phenomenon and will not be the last of its kind. First, coronavirus infection outbreaks have been common in the last two decades, despite previous episodes being relatively isolated at the regional level. As demonstrated by the first SARS outbreak in 2003 and the Middle East respiratory syndrome (MERS) outbreak in 2012, the zoonotic transmission of novel coronaviruses has resulted in severe threats to human health [17]. The COVID-19 pandemic was met with disparate responses in different countries, resulting in diverse impacts, with populations in Europe, the United States, and Latin America being disproportionately affected [18]. While science has learned a lot about SARS-CoV-2 and made incredible and remarkable progress in developing COVID-19 vaccines, there is still much uncertainty as the pandemic progresses. COVID-19 vaccines are being distributed in several countries, but this does not mean the crisis is over. We are now entering a new stage of the pandemic.

According to the current evidence, SARS-CoV-2, like SARS-CoV and MERS-CoV6, likely originated in bats and spread to humans through intermediate hosts. Coronavirus RaTG13 was discovered in the horseshoe bat *Rhinolophus affinis* in China's Yunnan province and was described as the SARS-CoV-2's closest relative. It has a 96.2% sequence identity with the SARS-CoV-2 genome, indicating that SARS-CoV-2 originated in bats [19]. Coronaviruses are composed of four genera belonging to the coronavirus family. This includes: alpha coronavirus, beta coronavirus, gamma coronavirus, and delta coronavirus. Unlike alpha coronaviruses and beta coronaviruses, which only infect mammals, gamma coronaviruses and delta coronaviruses infect a broader variety of animals, including birds, wherein SARS-CoV-2, SARS-CoV, and MERS-CoV belong to the beta coronavirus genus [20]. Despite a high degree of structural homology with SARS-CoV and MERS-CoV [21], SARS-CoV-2 is much more transmissible than its predecessors, which may be attributable to unique differences in the spike protein. Before the emergence of SARS, only two types of human coronavirus (HCoVs) were known. The first one was HCoV-229E, an alpha coronavirus that originated in bats and was transmitted to humans through alpacas. The second one was HCoV-OC43, a beta coronavirus passed from rodents to humans through cattle [22]. Two beta coronaviruses SARS-CoV and MERS-CoV, are unrelated, but both originated in bats and were transmitted to humans by wild carnivores and dromedary camels, respectively [23].

1.5.2 Some Animals Are Special: 12,000 Viruses like Coronavirus Are Waiting to Jump to Humans

Passage through wild carnivores and camels made CoV pathogenic, virulence increased manifold. Jumping between host species enables CoV to adapt to humans, but how these zoonotic agents like CoVs build that intelligence is unknown. The intelligence is not limited to CoV, but others like H5N1 avian influenza and SARS demonstrated zoonotic behavior. During the passage, the concept of the reservoir is important to model virus intelligence. Several novel CoVs with high genetic diversity from birds, including penguins, pigeons, peafowl, parrots, waterfowl, teal, quail, duck, and whooper swan (CSG) (CSG-https://talk.ictvonline.org/ictv-reports/ictv_9th_report/positive-sense-rna-viruses2011/w/posrna_viruses/222/coronaviridae). Apart from the bat, pangolin could be an intermediate animal reservoir before human transmission, causing COVID-19 (WHO-2020 situation report 22). As a result, zoonotic transmission of coronaviruses to human populations is unavoidable in the future. Understanding the origins and evolutions of coronaviruses and other viruses is crucial to understanding potential disease dynamics and developing informed strategies to prevent global dissemination.

1.6 Genomic Overview of SARS-CoV-2

SARS-CoV-2 is a single-stranded, positive-sense RNA virus with 16 non-structural proteins (NSP1-NSP16), four structural proteins (spike (S), membrane (M), envelope (E), and nucleocapsid (N)), and nine accessory proteins (Fig. 1.6; [24]). Coronavirus has one of the largest genomes of any RNA virus with a 27–32 kb genome and ~120 nm in diameter. Spike proteins on the virion surface are responsible for attacking host cells and causing viral and host cell membranes to fuse, both of which are crucial steps in initiating infection and allowing viral RNA to enter host cells. The virus shape and assembly/budding are regulated by membrane and envelope proteins, respectively. The SARS-CoV-2 genetic material reaches the host cell with

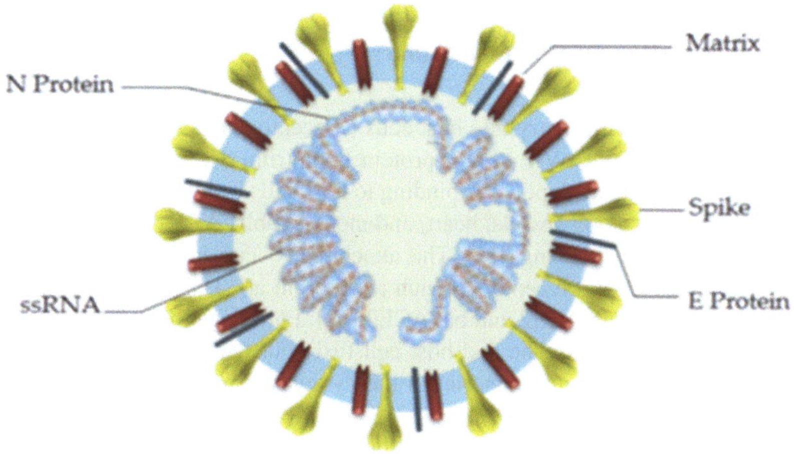

Fig. 1.6 Structure of severe acute respiratory syndrome coronavirus 2 [31]

the nucleocapsid protein, which aids RNA transcription, replication, virus assembly, and release [25, 26]. The spike, or S- glycoprotein, is a transmembrane protein in the outer portion of the virus with a molecular weight of around 150 kDa. The spike protein contains 1273 amino acids and is composed of two subunits, S1 (amino acids 14–685 and S2 (amino acids 686–1273, which are responsible for receptor binding and membrane fusion with the host cell, respectively, preceded by a short signal peptide (amino acids 1–13. The S1 subunit consists of three domains: an N-terminal domain (NTD, amino acids 14–305, a receptor-binding domain (RBD; amino acids 319–541, and a carboxy-terminal domain (CTD which has two subdomains (SD1 and SD2 [27, 28]. The S2 subunit consists of a fusion peptide (FP; amino acids 788–806) composed of hydrophobic residues, heptapeptide repeat 1 (HR1; amino acids 912–984), heptapeptide repeat 2 (HR2; amino acids 1163–1213), a transmembrane domain (TM; amino acids 1213–1237), and a cytoplasmic domain (CP; amino acids 1237–1273) [29, 30].

The nucleocapsid, also known as N protein, is the structural component of CoV, which is found in the endoplasmic reticulum-Golgi region and is structurally bound to the virus's nucleic acid content. The protein is involved in processes linked to the viral genome, the viral replication cycle, and the cellular response of host cells to viral infections since it is bound to RNA [32]. N protein is also highly phosphorylated, which is thought to cause structural changes that improve the protein's affinity for viral RNA. The membrane or M protein, which is the most structurally organized protein and plays a role in deciding the form of the virus envelope, is another essential component of this virus. This protein can bind to all structural proteins. Binding with M protein aids in the stabilization of nucleocapsids or N proteins and facilitates viral assembly completion by stabilizing the N protein–RNA complex within the internal virion. The envelope or E protein, the smallest protein in the SARS-CoV structure, plays a role in the virus's development and maturation [31].

1.6.1 Virus-Receptor Interactions

The key method of virus entry into host cells is receptor-mediated endocytosis. For viral infection, SARS-CoV-2 uses S-protein homotrimers that protrude from the viral surface and aid envelope virus binding to host cells by attracting ACE2, which is found in kidney, blood vessels, heart, and most notably, AT2 alveolar respiratory tract epithelial cells in the lungs [33]. This attachment occurs in the S-protein binding domain of SARS-CoV-2 receptors, which are present at 331–524 residues and can bind strongly to human ACE2 and bat ACE2 [32]. Following the entry and binding processes, the viral membrane and host cell fuse. Further to that, entered-SARS-CoV-2 will release genomic material in the cytoplasm and be translated into the nuclei.

The spike protein is found as a trimer on the virion's surface in its natural state, with the S1 and S2 subunits forming the extracellular stalk and bulbous "crown," which is translated as "corona" in Latin [28, 30, 34]. S1 subunit contains two subdomains: an

N-terminal domain and the C-terminal domain, known as a receptor-binding domain (RBD). SARS-CoV-2 uses its RBD to directly bind to the peptidase domain (PD) of ACE2. The trimeric spike protein's crown goes through hinge-like conformational shifts between a closed/down and a less stable open/up conformation [28, 30, 34, 35]. The RBD is accessible for binding to the ACE2 receptor in the open/up conformation, however, the RBD cannot interact with the ACE2 receptor in the closed/down conformation [36, 35]. The spike protein trimer undergoes a conformational change as it binds to the ACE2 receptor, making the S1 and S2 cleavage sites accessible to host proteases [37, 38, 39]. Cleavage of the S1 and S2 subunits prepares the spike protein for membrane fusion by allowing the S2 FP domain to be inserted into the host cell membrane, allowing the HR1 and HR2 coiled-coil domains to interact and form a six-helical bundle (6-HB). This bundle stabilizes another S2 subunit conformational shift that causes membrane fusion between the viral and host membranes [28, 30]. Coronaviruses express and replicate their genomic RNA during the intracellular life cycle to produce full-length copies incorporated into newly produced viral particles.

The spike (S) glycoprotein of coronaviruses is the most important viral protein for cross-species transmission and infection. It recognizes the host cellular receptor and mediates virus–cell membrane fusion. Cryogenic electron microscopy (cryo-EM) has recently been used to study the structure of the S-protein and ACE2 when they are bound to one of its usual ligands in the viral fusion [35]. Host proteases cleave S-protein at the S1/S2 boundary and the so-called S2′ site, located near the S2. Membrane fusion is facilitated by large irreversible conformational changes caused by this cleavage. The spikes of bat (RaTG13) and pangolin (PCoV GX) coronaviruses, which are closely related to SARS-CoV-2, were recently studied using cryo-EM [40]. These two spike trimers have all three receptor-binding domains (RBDs) in the "down" conformation, suggesting they are more likely to follow the receptor-binding inactive state. However, they discovered that the PCoV GX spike is comparable to the SARS-CoV-2 spike in binding the human ACE2 receptor and promoting pseudovirus cell entry, but not the RaTG13 spike. They also discovered essential RBD residues that underpin the RaTG13 and PCoV GX/SARS-CoV-2 spikes' separate activities.

1.6.2 Ongoing Evolution of SARS-CoV-2

The concern of mutations looms large nearly a year after the global Covid-19 pandemic began. It has enabled researchers to identify nearly 300,000 different SARS-CoV-2 genetic variants and link them to clinical and epidemiological findings. The answer to what will happen in the future is buried among this massive data collection, which constitutes the world's largest evolutionary experiment. Would mutation result in the emergence of new strains that are more infectious or immune to vaccines than the existing ones? New variants with the ability to spread faster are emerging, prompting the unavoidable question of whether they could render the recently approved vaccines ineffective.

On a molecular level, there is an impressive descriptive knowledge of all of these mutants. Several groups have clarified the relationship between the genetic sequence, the protein coded, and its role in the infectivity process, so its possible infectivity can be inferred [41–44].

The origin of new strains, such as B117 (also known as VUI202012/01), is extremely surprising. This latest strain has accumulated 23 separate mutations, 17 of which occurred all at once. Many spike protein mutations (e.g., H69 and V70, deletion 145, N501Y, A570D, D614G, P681H, T716I, S982A, D1118H) increase infectivity compared to the ancestral strain. This mutant form has been prevalent in Southern England, UK and is now spreading worldwide. Newly discovered mutant SARS-CoV-2 indeed have substantially higher biological effectiveness than the ancestral strains from which they originate. These different mutations would make it more difficult for the immune system to recognize the virus, leading to more patients contracting the disease twice and possibly necessitating changes to vaccines. To date, there is no proof that they are, but scientists are still investigating how the Covid-19 virus could evolve in the future and whether they can prevent it.

Polymerase chain reaction (PCR) tests, which look for traces of the virus's genetic material in samples, have been the key diagnostic method for Covid-19. To confirm the existence of an infection, it usually targets three sections of the virus. However, in samples from England with rapidly growing case numbers, one of these targets was progressively coming back negative, while the other two targets in the tests continued to work. When the scientists delved deeper, they found that the virus in these samples had picked up a mutation—a deletion at the same H69 and V70 positions in the protein. They discovered 16 other mutations that had changed the viral proteins they coded for, including some on the spike protein, in addition to this genetic alteration. They had found a new Covid-19 virus lineage that had acquired several mutations in a relatively short time. In the confusing world of coronavirus nomenclature, they called it B117—the new British Covid-19 variant, also known as VOC 202,012/01. By mid-January, it had carved a swath through the United Kingdom and had spread to 50 other nations. This recurrence of the same mutations in various variants offers some insight into what is going on: the virus could be subjected to similar evolutionary pressures as it spreads across millions of people, causing it to adapt in specific ways. Alone, these mutations are likely to have a minor effect. However, together, these different mutations would make it more difficult for the immune system to recognize the virus, leading to more patients contracting the disease twice and possibly necessitating vaccine changes. This site, known as a furin cleavage at the S1/S2 boundary site, enabled the virus to hijack an important enzyme in the human body. The SARS-CoV-2 S-mediated entry into cells was impaired by removing this cleavage motif [34]. At this stage, the enzyme snips the spike protein open, revealing secret sequences that help it bind more tightly to cells in the human respiratory tract, among other places. According to the researchers, a mutation close to this site could change this behavior even more. As a result, unique types of the SARS-CoV-2 cleavage site can be considered one of the factors affecting the virus's high rate of spread and pathogenesis [45].

Another study investigated the effect of SARS-CoV S-protein stabilizing mutations on S-protein conformational transitions [46]. They found that ACE2-binding or tyrosine cleavage at the S1/S2 boundary did not cause a significant conformation shift in the SARS-CoV S-protein stabilizing mutation. As a result, studies to develop effective antiviral drugs will concentrate on the furin-like cleavage that has not been found in other SARS-like CoVs [47]. Heparan sulfate, for example, was found to improve the open conformation of the RBD and thus the binding to ACE2 in a docking study [48]. S-protein is highly glycosylated, according to structural studies, which may influence receptor-binding kinetics. The structural dynamics of the S-protein are an inherent property, and the overall chemical, physical, and electrical charge of these carbohydrates have a significant impact on its characteristics.

Most random mutations during virus replication in the host cell do not affect the virus structure. However, sometimes a mutation changes the local conformation of certain functional groups in the spikes. The change could assist the virus that helps it enter cells, enter a cell of a typical organ, or avoid antibodies produced by the immune system. If an altered virus multiplies more quickly in the animal body or survives longer in the environment, or binds with wide ranges of cells, then it would spread more quickly and replace the parental form. The viral replicas of CoVs (an RNA-dependent-RNA polymerase) cannot read nucleotides properly during the replication cycle. The weakness becomes a boon. More it differentiates (once in 30 min), more mutations deviate viral progeny from the parental strain. This is called antigenic drift (Influenza also has this feature); it increases viral fitness in different animals. There is another interesting feature of CoVs. Upstream each gene of a CoV RNA, there is a unique consensus sequence of genes. They not only bind with various CoV strains but also other viruses too. If there is coinfection, such recombination enables RNA polymerase to learn rapidly other strain's mutations, virulence, host features, i.e., pathogenicity) [49].

Before the emergence of SARS, only two types of human coronavirus (HCoVs) were known. The first one was HCoV-229E, an alpha coronavirus that originated in bats and was transmitted to humans through alpacas. The second one was HCoV-OC43, a beta coronavirus passed from rodents to humans through cattle [22]. Two unrelated beta coronaviruses SARS-CoV and MERS-CoV, but both come from bats and transmit to humans by wild carnivores and dromedary camels, respectively [23]. Passage through wild carnivores and camel made CoV pathogenic, virulence increased manifold. Jumping between host species enables CoV to adapt to humans, but how these zoonotic agents like CoVs build that intelligence is unknown. The intelligence is not limited to CoV, but others like H5N1 avian influenza and SARS demonstrated zoonotic behavior. During the passage, the concept of the reservoir is important to model virus intelligence. Several novel CoVs with high genetic diversity from birds, including penguins, pigeons, peafowl, parrots, waterfowl, teal, quail, duck, and whooper swan (CSG). Apart from the bat, pangolin could be an intermediate animal reservoir before human transmission, causing COVID-19 (WHO-2020 situation report 22). Moreover, when the vaccine is used uncontrollably, field viruses and partially immune birds catalyze recombination between vaccine molecule and the virus strain leading to high genetic variability [50–52]. Variability caused by any reason does not remain limited to any particular species; it spreads beyond that species.

As a result of its propensity for mutation and recombination, the S-protein of SARS-CoV-2 continues to diversify. During the pandemic, analyses of globally sampled SARS-CoV-2 genomes indicated the existence of many subgroups of the virus with distinct mutations. At least 17 major SARS-CoV-2 lineages have been identified to date, each with many distinct single nucleotide polymorphisms (SNPs) [53]. Understanding the practical ramifications of these SNPs has huge implications for predicting the pandemic's possible trajectories and designing prevention and treatment strategies. Furthermore, widespread human infection is now posing additional threats to other mammals. Secondary and even tertiary transmissions between humans and other mammals can occur in species that communicate with human populations. The emergence of potentially harmful recombinant SARS-CoV-2 strains could be aided by such cross-specific transmission. To decide how to slow down and eventually eliminate intra- and between-species transmissions, epidemiological, genetic, and functional studies of variants must be completely exploited.

The S-protein mediates viral attachment to specific cell receptors, and it is the main inducer of virus-neutralizing antibodies. The SARS-CoV-2's RNA genome is 82% identical to the earlier SARS-CoV with some viral proteins ~90% homologous to SARS-CoV [54]. Five of the six receptor-binding domain (RBD) or amino acids residues of SARS-CoV-2 are optimized for binding to the human receptor ACE2 compared to SARS-CoV-like viruses. For SARS-CoV, the RBDs are Y442, L472, N479, D480, T487, and Y4911, corresponding to L455, F486, Q493, S494, N501, and Y505 in the SARS-CoV-2. Five of these six residues changed during evolution, such that the spike binds efficiently with any species with high receptor homology [55]. Moreover, the spike glycoprotein of SARS-CoV-2 has a functional polybasic (furin) cleavage site at the S1(trimetric front, receptor-binding domain)–S2 (cylindrical tube with an RNA sequence) boundary of the spike through the insertion of 12 nucleotides. The polybasic cleavage sites and the three adjacent O-linked glycans are unique to SARS-CoV-2 and were not previously seen in the lineage of B beta coronaviruses. S-glycoprotein is the key, it must be cleaved if a virus at all infects a cell, but it varies widely across animals.

CoVs are ~120 nm in diameter, enveloped with 20 nm glycoprotein spike, three structural (envelope) proteins, for example, membrane (M), envelope (E), and spike (S) are coordinated with a single positive-strand RNA genome of approximately 30 kilobases (kb) in length. The S-protein mediates viral attachment to specific cell receptors, and it is the main inducer of virus-neutralizing antibodies. The SARS-CoV-2's RNA genome is 82% identical to the earlier SARS-CoV with some viral proteins ~90% homologous to SARS-CoV [54]. Five of the six receptor-binding domain (RBD) or amino acids residues of SARS-CoV-2 are optimized for binding to the human receptor ACE2 compared to SARS-CoV-like viruses. For SARS-CoV, the RBDs are Y442, L472, N479, D480, T487, and Y4911, corresponding to L455, F486, Q493, S494, N501, and Y505 in the SARS-CoV-2. Five of these six residues changed during evolution, such that the spike binds efficiently with any species with high receptor homology [55]. Moreover, the spike glycoprotein of SARS-CoV-2 has a functional polybasic (furin) cleavage site at the S1(trimetric front, receptor-binding domain)–S2 (cylindrical tube with an RNA sequence) boundary of the spike through the insertion of 12 nucleotides. The polybasic cleavage sites and the three adjacent O-linked glycans are unique to SARS-CoV-2 and were not previously seen in the lineage of B beta coronaviruses. S-glycoprotein is the key, it must be cleaved if a virus at all infects a cell, but it varies widely across animals.

1.7 A Composition of Cavity and Dielectric Resonators: Virus–Receptor Complexes

The novel human coronavirus called severe acute respiratory syndrome coronavirus 2 (SARS-CoV-2) is testing human immunity globally. The debates are mounting on how the virus develops intelligence that perfected the binding of its corona-like spikes with the human receptor cells. Here we report a generic model to understand the intelligence of virus evolution. By studying the electromagnetic and dielectric resonance of the virus structures bonded to the ACE-2 receptor cells of various animals and humans, we found that for the last 18 years, the SARS virus has been on a mission to perfect its bonding with the ACE-2 receptor cells. We tracked how zoonotic assisted complete SARS-ACE-2 bonding and we found a universal evolutionary route. We have predicted how SARS-Cov-3 and SARS-Cov-4 will look when they arrive in the near or distant future. While developing the generalized evolution path for SARS-Cov-2, we used our universal intelligence model to build an artificial brain, namely, Phase Prime Metric (PPM). Our model paves the way to build a real-time monitor for any virus evolving similarly, silently.

Here we have looked into the RBD binding domains of SARS-CoV-2 and cell receptor ACE-2 for the strains available from different species. In this work, we study structures after cleavage and docked with the receptor. We collected the virus–cell docking structures from the PDB database and then fed those structures into the (Computer Simulation technology-CST) computer simulator. In the simulator, we put ports for electromagnetic energy minimization and solve Maxwell's equation to determine how electromagnetic energy transmits through the junction. Specifically, we measure reflection and transmission coefficients as a function of applied frequency. At resonance frequencies, we determine potential distributions on different parts of the biological structure (Table 1.1). For dielectric resonance simulation, we modeled the entire virus spike–receptor complex as a composition of the cavity and dielectric resonators. For example, we have studied seven structures: (1) 3sci SARC coronavirus complex with the human receptor, (2) 3scl human chevet corona-ACE2 complex, (3) 4Kr0 virus accepting cell surface protein CD26 with the virus, (4) 4l72 typical human cell receptor and bat coronavirus complex, (5) qzv bat-derived coronavirus human cell entry complex, (6) 6lzg latest work human receptor ACE2 and coronavirus binding, and (7) 6m0j similar to 6lzg as shown in column II of Table 1.1, respectively.

Table 1.1 Electromagnetic distribution profile of various protein complexes at their resonance frequency on the 2D cross-section plane

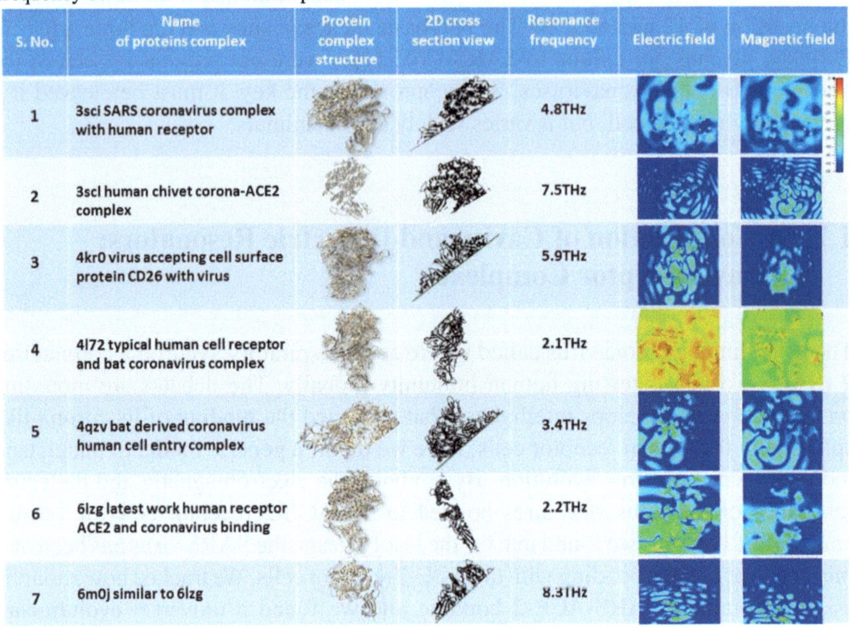

S. No.	Name of proteins complex	Protein complex structure	2D cross section view	Resonance frequency	Electric field	Magnetic field
1	3sci SARS coronavirus complex with human receptor			4.8THz		
2	3scl human chivet corona-ACE2 complex			7.5THz		
3	4kr0 virus accepting cell surface protein CD26 with virus			5.9THz		
4	4l72 typical human cell receptor and bat coronavirus complex			2.1THz		
5	4qzv bat derived coronavirus human cell entry complex			3.4THz		
6	6lzg latest work human receptor ACE2 and coronavirus binding			2.2THz		
7	6m0j similar to 6lzg			8.3THz		

Similarly, once we receive the molecular structure, be it a virus or a virus–cell receptor, we painstakingly recreated all secondary structures of the protein components one by one [56]. The artificial protein structure was created with a similar size, a similar shape, and a similar dielectric property in the CST simulator. While simulating the objects to solve Maxwell's equations, find the frequencies at which transmission and reflection of electromagnetic signals are maximum, i.e., resonance frequencies. We do not claim that electromagnetic energy transmission is causing the Coronavirus. Our study is limited to the 5–8 THz domain, where there is a possibility that thermal noise is already available in a 37 °C human body or most living life forms. Moreover, electromagnetic resonance is used for medical treatment [57].

1.7.1 Potential Distribution at the Protein Complex Junction

EM energy reflects the deposition wave nature at the junction of protein complexes. Table 1.1 defines the virus–receptor junctions of protein complexes in the sense of electromagnetic energy. Many viruses belong to the coronavirus family and can infect a high range of host/receptor cells, including humans, birds, and other mammals. Different virus cells form strong junctions with different receptor cells, as listed in column II of the table. Here we have reported electromagnetic energy profiles at the junction of coronaviruses of different species with human receptor cells. Here the protein complexes are nanometer-sized and have curved surfaces. Electromagnetic energy is applied through the waveguide source. Somewhere the electric vector is absorbed more and somewhere, magnetic vector. The electric and magnetic part coupled to the electromagnetic wave defines some undefined field where the complementary part bursts for energy conversion, called an evanescent wave. Using a simulation study, we found that electromagnetic energy is the companion to the evanescent waves (an oscillating nature of the electric/magnetic field) that is not propagated like the electromagnetic field but is concentrated near the energy source. The electric and magnetic field part is highly decoupled in the evanescent wave [58]. It may be a pure electric vector or magnetic vector, or a mixture of both vectors. In columns VI and VII of the table, the ripple of the electromagnetic wave has the key information of the proteins' complex geometry and wave decay exponentially. The human receptors with the bat coronavirus complexes have high energy intensity, whereas the SARC coronavirus with the human receptor has a complex ripple form. Energy intensity magnitude matters the absorption of electric and magnetic vectors through the protein surface. The electromagnetic energy is transmitted to the junction, and it drops off exponentially from the virus to the receptor junction.

Several phylogenetic tree-based analyses have been carried out to understand the relationship between SARC-CoV-2 with other Coronavirus species. A previous study suggested that the genomic sequence of SARS-CoV-2 is similar to BAT-CoV. A comparative study is available on MERS_CoV, SARC-CoV, and BAT-CoV [59]. Rehman, H. A. et al. identified the genetic markers like mutations in proteins, SNPs (single nucleotide polymorphisms), and whole-genome sequence phylogeny.

1.7.2 How Does Our Study Differs from the Existing Studies?

Our study is unique for two reasons. We have not looked into the 2D vertical cross-section for the potential distribution of the protein junction as it is done in conventional literature. Instead, we have taken a 2D horizontally flat surface area between the spike protein of SARS-CoV-2 and the human receptor cell ACE-2 (angiotensin-converting enzyme-2). The reason for choosing a 2D horizontal flat surface instead of a commonly used 2D vertical area for developing a potential map is that we found from the crystallography structures of various spike-ace junctions that four contact points are changing potential simultaneously (Figs. 1.7a–c and 1.8). Most research papers on virus–human cell interactions concentrate on the vertical cross-section at the virus–human cell junction. As a result, they find only the local 2D changes in the H bonding and dipole–dipole interaction between the spike and the ace-2 virus. Based on changes in the bond length for the junctions of spike-ace2 among different species, the evolution process of the virus is analyzed. Here we suggest looking orthogonally rather than the conventional route. The junction profile undergoes a major change, and several unique features become visible.

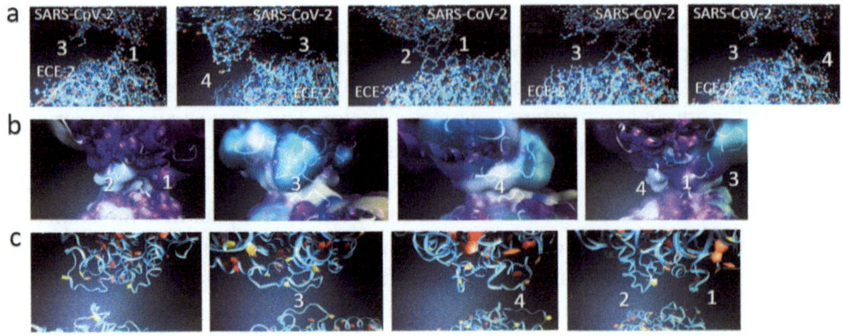

Fig. 1.7 **a** PDB database code 3SCI SARS-CoV-2 Coronavirus (top) and ace-2 (bottom) complex. We have rotated the junction 360° and put numbers 1–4 to show four possible junctions that we observed for various species. **b** The same structure with the hydration shell so that the interactive domain is explicit. **c** Benzene rings are brightened; it helps in identifying the residues connected to a particular site

We have theoretically simulated 18 junctions of SARS-Cov-2 and human ace-2 cell receptors. Some of the structures are already available in the PDB database. SARS-Cov-3 and SARS-Cov-4 are artificially created. We have tried multiple combinations of available virus spikes with human receptors. At the junction, we noted the potential distribution if nW of energy is pumped externally, whose origin in a real cellular environment could be the 5–6 THz thermal noise (kT). We have fitted 2D surface potential patterns with the geometric shapes of our already proposed and patented GML and PPM models for predicting the change of symmetries or geometric structures. Thus, we concluded a possible geometric shape if the virus continues to evolve the junction similarly. The difference with other models is that we do not use any assumption or fit output with an equation. Change in geometric shapes follow a rule, and that is implemented here.

1.7.3 Future Evolution Path of SARS-CoV-2

When coronavirus transmits from one person to another person, a tiny change occurs in its genetic code. However, researchers are starting to notice the patterns during which the virus is mutating. When a patient catches the coronavirus, his immune system goes into a bad situation. Convalescent plasma (milky-brown liquid) is an antibody against the virus which neutralizes it. Researchers looked at some remarkable changes in the virus's genetic material over time. They suggested that the virus shows the sign of adaption to avoid the antibodies in the convalescent plasma. Two amino acids like H69 and V70 in the spike protein play a key role in the ability of Coronavirus to infect the host cells. Spikes are the main way to recognize the cells infected by the covid-19 virus and help the virus penetrate the cells. The mutation of the H69/V70 increases the infectivity, which prompted researchers to generate

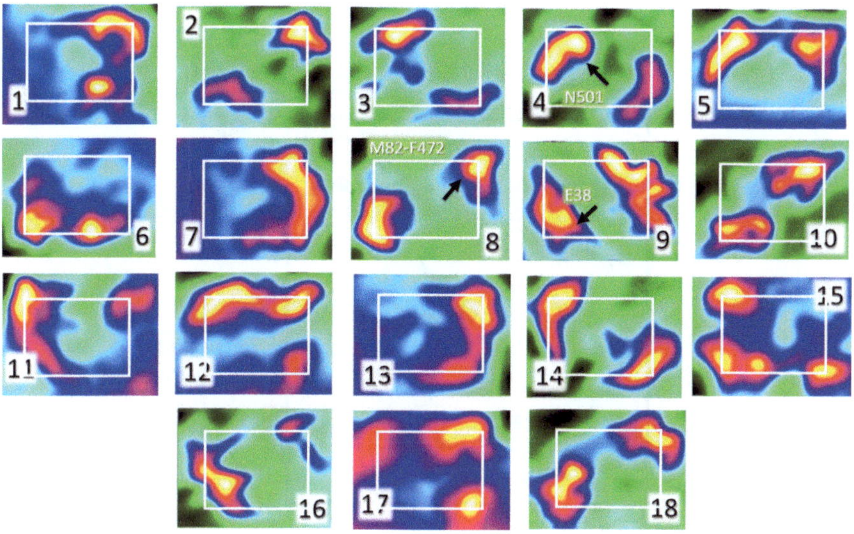

Fig. 1.8 Normalized potential distribution (eV) for at the virus–receptor junction 18 structures (0.2 nm × 0.2 nm). A square is drawn to show the virtual geometric structure around which the bonding locations evolve across species. Here is a list of basic structures that we adopted from the free PDB database (note that mice and rats are absent because their ace-2 is not recognized) and estimated potential at the horizontal cross-section, using molecular simulation. The same structures run on CST, but one has to connect two ports. Here we note the PDB file name if readers wish to study further. 1. 1r42 pure human ACE2 structure with SARS-Cov-2; 2. 2bat influenza cell receptor with SARS-Cov-2; 3. 3kbh coronavirus respiratory cell receptor with SARS-Cov-2; 4. 3sci SARS coronavirus complex with human receptor; 5. 3scl human civet corona-ACE2 complex; 6. 4kr0 virus accepting cell surface protein CD26 with a virus; 7. 4l3n MERS coronavirus binding spike protein; 8. 4l72 typical human cell receptor and bat coronavirus complex; 9. 4qzv bat-derived coronavirus human cell entry complex; 10. 5gyq bat coronavirus pure NOT complex; 11. 5wrg Typical conformation required for easy entry of SARS-CoV in the human cell; 12. 5×4 s N-terminal for bonding SARS-CoV; 13. 6l8q interesting bat CD26 and MERS RBD complex; 14. 6lzg latest work human receptor ACE2 and coronavirus binding; 15. 6m0j similar to 6lzg 3sci and 3scl corona and ACE2 but subtle difference is found if we rotate; 16. 6vsb 2020 coronavirus binding human cell; 17. 6vw1 coronavirus ACE2 complex. 18. 6mwm (bat coronavirus) + 5gyq (bat receptor). If not mentioned, then we use the human ace2 receptor or human SARS-Cov-2 by default

an international genetic database of Covid-19. Researchers are trying to understand how the covid-19 virus evolves in the future how it could be stopped.

One of the foremost observations on the horizontal cross-section was the four sites, two from the spike and two from the receptor, change their relative positions during evolution. As a result, when they bind with each other, depending on the relative positions of the nucleic acid residues, the iso-potential contour changes around four points of a virtual square (Fig. 1.9). We have tracked the horizontal cross-sections for various species, for the SARS-CoV-2 and human cell receptor ACE-2 junctions. The most important observation we have made is that as time passes by, the CoV strain has perfected the locking arrangement with the receptor. The center of the virtual

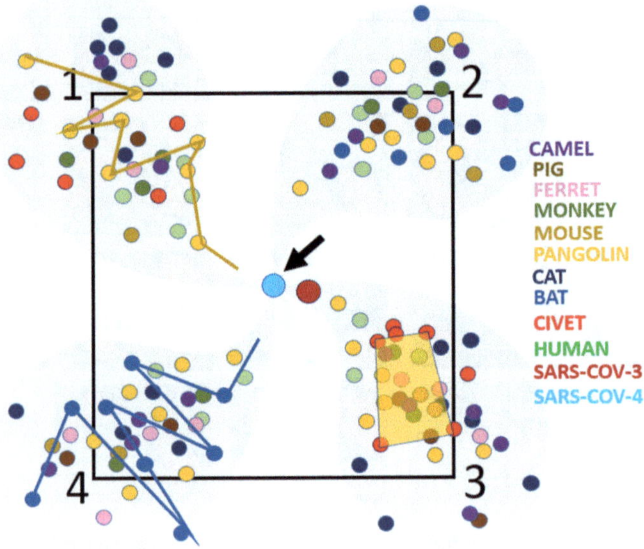

Fig. 1.9 The centers for key potential distributions for different species, SARS-human 2002, 2008; SARS-civet; SARS-bat 2013; we artificially changed ace2 for mouse, rat pig, ferret, cat, monkey, Dromedary camel coronavirus HKU23 (6JHY), and pangolin (PDB code 6ACD). Geometric information for constructing time crystal and the predicted SARS-Cov-3 and SARS-Cov-4 structures (further RBD mutation) along with SARS-Cov-2 with ACE-2 complex (PDB ID 2AJF)

square moves to the ideal location. If we extrapolate the path we have noticed thus far, we could predict where the center would be when SARS-CoV-3 and SARS-CoV-4 would come (Fig. 1.9). In summary, the junction has made an effort to position itself in the energy minima of the potential well created by the two dynamic domains, the CoV spike, and the ace-2 receptor.

During the simulation of extrapolating the path for the future evolution of SARS-CoV-2, we find that complex geometric shapes of various pathways are slowly converged to the smaller shapes (Fig. 1.9). Decreasing entropy via simplifying topology is normal. However, the domination of particular species in a particular domain of the virtual square is unique. We find that certain geometric pathways are followed in the particular domain around the square dedicated to a particular species. For multiple species, particular geometric pathways are noted. This is expected since protein structures are not random across species, the conformations of the secondary structures remain similar. A local shift of a nucleic acid center by 0.2–0.3 nm is considered a major change; our previous studies on molecular conformers have intricately documented such features [60, 61]. For that reason, primarily, one could see that species-specific topological transformations are obvious in the virtual square that we consider a generic topology for CoV evolution across the species.

Finally, we calculate the symmetry of the geometric pathways adopted by the species on the virtual square and find that the trial and error process chooses symmetry as the key search engine (Fig. 1.9). It means if, at the initial stage of the search process, a triangle is made of 2:3:4, then during random changes in the structure as part of virus evolution, the junction would undergo changes causing significant changes in the geometric shape. The ratio of sides 2:3:4 could switch to 2:4:3 or even 3:2:4. Such an effort was observed earlier in the vibrational frequency response of the microtubule at resonance. During the electromagnetic resonance measurement of the biomaterials, it was observed that the ratio of frequencies shift as if one is calculating the ordered factor of integers [62, 63, 64, 65]. Based on such observations, the idea of a phase prime metric was conceived, which is primarily a catalog of all possible compositions of symmetries.

1.8 Basic Understanding of Generic Model for Virus Evolution Using the Time Polycrystal

CoV uses a square to evolve various species; that does not mean the other virus would do the same. The virtual geometric shape generated at the horizontal junction of the virus and the receptor cell could become a triangle or pentagon. The geometric musical language, GML, and the phase prime metric, PPM, are two already established models that operate together. We have demonstrated GML-PPM-based intelligence in artificial protein-like molecular systems [66, 67]. GML accounts for vibrations and converts them into a time crystal, while PPM converts time crystal into a complex geometric structure considering all possible symmetries. Here is a process of building a generic model for the virus evolution using the composition of time crystals or time polycrystals. Both Winfree and Wilczek versions of time crystals were limited to only one clock, they never had to be concerned with the nature and features of the singularity domain, but here we have to do that; hence UTC (Universal time crystal) is essential [68]. For a virus, time polycrystal means a composition of geometric pathways that a particular species adopts during its mutation and consequent evolution process.

1.8.1 The Origin of Intelligence in a Virus: How Do They Outperform Human Immunity?

The universal quest to learn how a simple life form, e.g., a virus made of only a few protein molecules, runs by itself, evolves, and outperforms human immunity gave birth to the concept of a time crystal in the 1970s [69]. When the biological clocks are disturbed, chemically, physically, or otherwise, they return following a unique path along the return route [70, 71]. It creates new clocks. When the system returns periodically post perturbation, the phase plots of the intermediate new clock look like a crystal, so it was called a time crystal in the 1970s [69]. Virus self-assembles in magic numbers and forms platonic solids, i.e., elementary physical crystals, simply by entropy minimization and solving a fundamental mathematical problem, how to pack circles on a sphere [72]. The clock-like properties of biological time crystal research were buried until we universalized it and used it for developing intelligent machines of the future [73–76]. We have developed a metric of primes and all possible compositions to build a temple-like structure that accommodates 86% of all possible breaking symmetries [68]. This mathematical structure is named the phase prime metric (PPM). We used it earlier to build an artificial brain [76], and here we would use it to build a structure that delivers intelligence to the SARS-Cov-2 virus. The fundamental conceptual change would be that the vibrational frequencies and evolutionary changes of viruses would be geometric, platonic, or Archimedean [77] when we would see the evolution of the virus in terms of time crystals.

1.8.2 Applying Geometric Shapes to Our Already Reported Space-Time-Topology-Prime (STtS) Metric

Normally a metric measures the spatial distance, for self-assembly of time polycrystals, which is the information processing unit in the brain and the virus and its entire network. Here we have considered the evolution of entire living species in terms of singular genetic code, considering that there was a primitive genetic code that evolved over time to create the current living kingdom on our planet. The entire living kingdom and its evolution is regulated by a metric that is independent of the local genetic information of a particular species. The polycrystal's growth within and above the virtual square for CoV is represented by three projection-feedback (PF) networks that could be measured using the space $ds^2(h_3)$, time $dt^2(h_{1,2})$, topology $dT^2(h_5)$, and symmetry $dS^2(h_7)$. Here the combinations of dual, triple, and quad features are:

When a time crystal is formed at the RNA replication level, we get quaternions that could have information like space-time $st(h_3, h_{1,2})$, with 3 + 1/2 = 5/6 dimensions. A small number of RNA codes that are inserted into the cell through the CoV spike, generate full code, we estimate them using octonions, we could have information of Space-topology $sT(h_3, h_5)$, with 3 + 5 = 8 dimensions; time-symmetry $tS(h_{1,2}, h_7)$, with ½ + 7 = 8/9 dimensions. Finally, at the highest level during recombination of similar virus strain RNAs or other viruses (coinfection), dodecanion would represent the information content. Therein information type would be topology-symmetry $TS(h_5, h_7)$, with 5 + 7 = 12 dimensions; $stST(h_3, h_{1,2}, h_5, h_7)$, with 3 + 1/2 + 5 + 7 = 16 or 17 dimensions. Space-symmetry $sS(h_3, h_7)$, with 3 + 7 = 10 dimensions; Space-topology $tT(h_{1,2}, h_5)$, with ½ + 5 = 6/7 dimensions; space-symmetry-time $sSt(h_3, h_7, h_{1,2})$ with 3 + 7 + 1/2 = 11/12 dimensions;

space-time-topology $stT(h_3, h_{1,2}, h_5)$ with 2 + 1/2 + 5 = 8/9 dimensions; space-symmetry-topology $sST(h_3, h_7, h_5)$ with 3 + 7 + 5 = 15 dimensions; symmetry-time-topology $tST(h_{1,2}, h_7, h_5)$ with ½ + 7 + 5 = 13/14 dimensions; symmetry-space-time $sSt(h_3, h_7, h_{1,2})$ 3 + 7 + 1/2 = 11/12 dimensions. Two dimensions are taken into account when a new kind of dynamic feature is added to a new axis; we go up to 20D (icosanion). However, when we consider the dimension of a tensor representing a situation where the dynamics of one system are nested inside another, then the maximum is 12D (dodecanion) [78]. In three PF layers, all noted tensors listed above generate continuously generate various distinct dynamics represented as time polycrystals, and those virtual particle-like elements would interact with the phase prime metric, PPM $\Delta_{\text{PPMi}}(i = 1 \text{ to } \infty)$.

Hence, $stST$ would be confined to one species, i.e., it would represent the projected and the feedback time crystal "to and from" the PPM. The metric representing the time crystal in stST = $S2T2$ universe of 12 nested worlds is given by [79].

$$
\begin{aligned}
H = &\Delta_{\text{PPM1}} st(h_3, h_{1,2}) + \Delta_{\text{PPM2}} sT(h_3, h_5) + \Delta_{\text{PPM3}} tS(h_{1,2}, h_7) \\
&+ \Delta_{\text{PPM4}} TS(h_5, h_7) + \Delta_{\text{PPM5}} sS(h_3, h_7) + \Delta_{\text{PPM6}} tT(h_{1,2}, h_5) \\
&+ \Delta_{\text{PPM7}} sST(h_3, h_7, h_5) + \Delta_{\text{PPM8}} stT(h_3, h_{1,2}, h_5) + \Delta_{\text{PPM9}} tST(h_{1,2}, h_7, h_5) + \\
&+ \Delta_{\text{PPM10}} sSt(h_3, h_7, h_{1,2}) + \Delta_{\text{Project-Feedback}} stST(h_3, h_{1,2}, h_5, h_7)
\end{aligned}
$$

One has to map the junction horizontally for any two recombinations for any junction. Once the geometric shapes are mapped, topology and symmetry are determined. Finally, as soon as one gets the variations in terms of primes, H would deliver how much variation the virus evolution has produced and what would be the next in the evolution process. In Fig. 1.9 for SARS-Cov-2, the output dynamics are like a composition of four spirals with topological charge +2 or −2. The PPM-GML-based intelligence-based models are superior to the neural network and genetic algorithm-based models [80].

While building the spike–receptor, we observed that nature has repeatedly used a few geometric shapes, and these teardrops to ellipsoid transition could be represented using a single mathematical equation. We have also found that the prime number of cavities, resonating centers, or oscillators were assembled to build a junction between the virus and the receptor. Therefore, the intelligence of a virus is not limited to a virus of a single strain or a virus colony. It is the collective output of various hosts as all effects are integrated into a space-time-topology-symmetry (stTS) metric that provides 11D composition of tensors representing the geometric information structure of a self-operating mathematical universe. The reason for self-operation is that the pattern of primes called a phase prime metric (PPM) (see Fig. 1.10) that links all events represented as geometric shapes embedded in the time crystals [68], builds stereographic projection to infinity. The feedback from infinity returns the topological fraction of the projection to infinity. Thus, mostly the feedback is not defined in the set of stereographic projection to infinity [79]. Space-time-topology-symmetry (stTS)

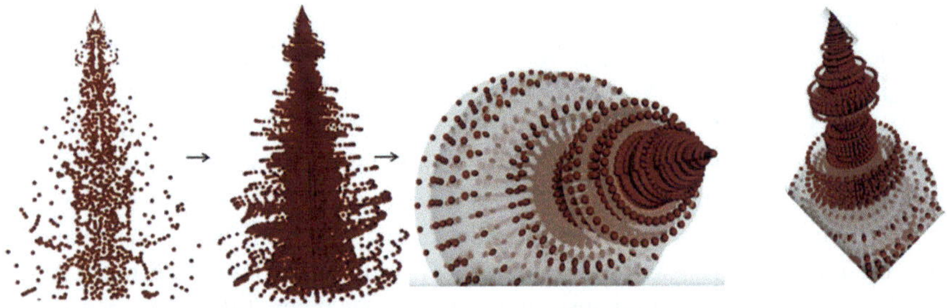

Fig. 1.10 We counted initially 15 primes and suggests 15 primes cover 99.99%. events of the universe. If we put integer on the x-axis and their contributions (contribution of 4 is 1(2 × 2); 5 has 0; 6 has 2 (2 × 3 and 3 × 2); etc.) on the y-axis and rotates whole arrangements at 360 degrees, we get phase prime matrix, PPM

metric encompasses the entire guest host system in the living network that participates in recombination.

We have introduced 12 imaginary words, dodecanion algebra from eight imaginary worlds, octonion algebra (Fig. 1.11). Dodecanion algebra is the 12 imaginary worlds one inside another. The only difference between 8 and 12 imaginary worlds is the fano plane. A set of imaginary vectors is replaced by manifold triplets, which may have three forms. Quaternion, octonion, dodecanion, and icosanion algebras are the composition of prime dimensional vectors. The stST, the space-time-topology-prime metric is made from the geometric shape as topology and primes as symmetry. stST is a self-operating mathematical universe with many numbers of nested imaginary worlds. We get the decision in the form of shape-changing geometry with time by stST.

1.8.3 Electromagnetic Interactions of Proteins and Time Crystals

Finally, proteins and nucleic acids are so designed that at electromagnetic resonance, the electric E and magnetic M parts of the electromagnetic field get a nonlinear treatment. In some regions, E is absorbed more; in some regions, the M vector is absorbed more. The result is the creation of an Evanescent wave [58, 81], which generates vortex or free atom-like structure of fields [82]. We measured the electromagnetic resonance of proteins and their complex structures inside neuron cells to determine relevance information processing [64, 66, 83]. The number of singularity points on the primary phase cycle is the number of guest clocks [84]. It is the number of different time flows experienced by a system point as it moves 360°. Each clock can have its system point and grow its phase cycle structure or time crystal inside by making a new guest–host system. It can connect with neighboring time crystals or phase cycles as guests of a larger phase cycle. The self-assembly of time crystals

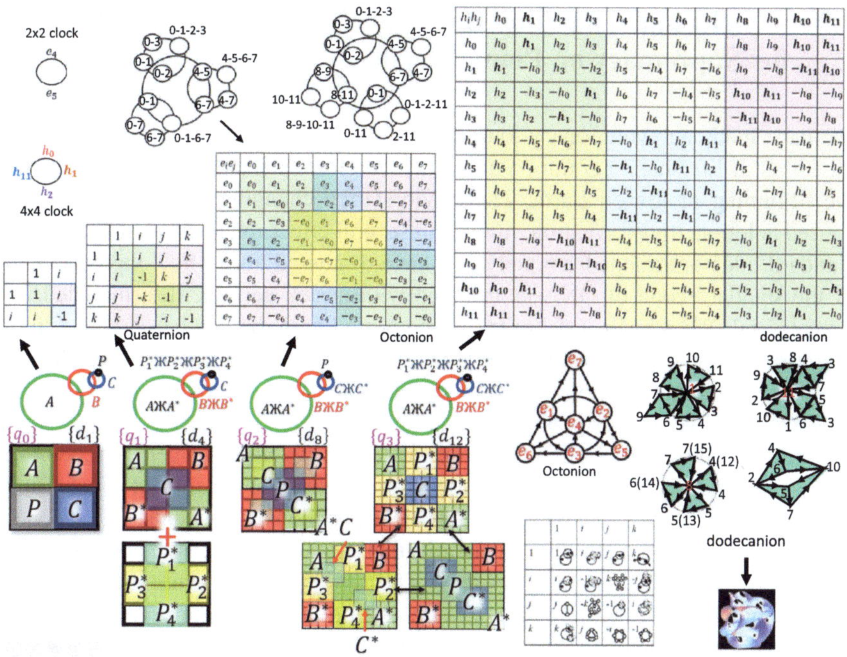

Fig. 1.11 Four types of tensors (dinion, quaternion, octonion, and dodecanion) are shown for time crystal representation. Multinions are explained in four different ways. First, time crystal, second tensors, third colored square matrix, and fourth linguistic representation. Tensors are the combination of a subset of each clock. Each clock is a submatrix of entire tensors like A, B, P, and C. Quaternion shows dual tensor, while dodecanion is the simultaneous activation of three tensors. The bottom right end of the figures shows the fano plane and manifolds for quaternion and dodecanions algebra. The clocking-like architecture of quaternion at the bottom right, which suggests a single element in tensor, seems like topologically

side by side [85, 86]. Therefore, when a vector of any dimension encounters a singularity point [87] during destructive interference of vortex atoms (nerve spike) around that point. The symmetry is normally written in terms of associative primes change, which is called prime shift. The origin of prime shift is that in situ updates of time crystals perpetually in the guest–host system of a virus. Moreover, the effect of information exchange, learning, and plasticity, decision-making, for any virus mutation and evolution, a change in the symmetry or creation or editing of singularity points is an inevitable process [68].

1.9 Conclusion: The Final Battle of Virus and Humans for Supremacy on Planet Earth

We have made an evolutionary journey of virus in this chapter. It is assumed that all forms of life share a common ancestor; therefore, a standard code was there at the beginning of life on Earth. Then code started changing; codon reassignment caused deletion, transformation, jump, and several effects on the standard code. The standard code is arranged uniquely; we have discussed above how the periodicity of the standard code of DNA is arranged with nested periodicities, modeled as a time crystal. We are not just limited to the nested periodicities but also looking at the resonant oscillations of these periods. DNA code holds a history of life form in our planet, and that history maps how we humans have reached the shape we hold now and what we would become.

There are three proposals on the origin of the evolution of genetic code. The first is the stereochemical theory. Following this theory, the physicochemical affinity between amino acids and cognate codons dictates codon assignments. The second is the coevolution theory. This theory suggests that as amino acid biosynthesis pathways evolve, the code structure changes with it. The third is the error minimization theory. According to this theory, mutation and translation errors are minimized and in the proccss of minimization, codon rcadjustment takes place.

Covid-19 has shown that it is not far that more than 12,000 viruses are waiting in the animals to jump to humans. The collective evolution of the standard codes of DNA and RNA would make humans more suitable to virus attacks so that over time humans will either undergo massive genetic transformations or disappear from the planet forever.

References

1. Koonin EV, Novozhilov AS (2009) Origin and evolution of the genetic code: the universal enigma. IUBMB Life 61(2):99–111. https://doi.org/10.1002/iub.146
2. Nelson MI, Holmes EC (2007) The evolution of epidemic influenza. Nat Rev Genet 8:196–205. https://doi.org/10.1038/nrg2053
3. Andersson SGE et al (1998) The genome sequence of *Rickettsia prowazekii* and the origin of mitochondria. Nature 396:133–143. https://doi.org/10.1038/24094
4. Koonin EV, Martin W (2005) On the origin of genomes and cells within inorganic compartments. Trends Genet 21:647–654
5. Prangishvili D, Forterre P, Garrett RA (2006) Viruses of the archaea: a unifying view. Nat Rev Microbiol 4:837–848. https://doi.org/10.1038/nrmicro1527
6. Brattas PL et al (2017) TRIM28 controls a gene regulatory network based on endogenous retroviruses in human neural progenitor cells. Cell Rep 18(1–3):1–11
7. Vallender EJ, Mekel-Bobrov N, Lahn BT (2008) Genetic basis of human brain evolution. Trends Neurosci 31:637–644
8. Hogue MJ et al (1955) The effect of poliomyelitis virus on human brain cells in tissue culture. J Exp Med 102(1):29–36

9. Paterson RW et al (2020) The emerging spectrum of COVID-19 neurology: clinical, radiological and laboratory findings. Brain 143(10):3104–3120. https://doi.org/10.1093/brain/awaa240
10. Ladecola C et al (2020) Effects of COVID-19 on the nervous system. Cell 183(1):16–27
11. Moriguchi T et al (2020) A first case of meningitis/encephalitis associated with SARS-Coronavirus-2. Int J Infect Dis 94:55–58
12. Zanin L et al (2020) SARS-CoV-2 can induce brain and spine demyelinating lesions. Acta Neurochir 162:1491–1494
13. Bryce C et al (2020). Preprint at medRxiv. https://doi.org/10.1101/2020.05.18.20099960
14. Song E et al (2020). Preprint at bioRxiv. https://doi.org/10.1101/2020.06.25.169946
15. Al Saiegh F et al (2020) Status of SARS-CoV-2 in cerebrospinal fluid of patients with COVID-19 and stroke. J Neurol Neurosurg Psychiatry 91:846–848
16. Chenga Q et al (2020) Infectivity of human coronavirus in the brain. E Bio Med 56:102799
17. Dudas G, Carvalho LM, Rambaut A, Bedford T (2018) MERS-CoV spillover at the camel-human interface. Elife 7:e31257
18. Brauner JM, Mindermann S, Sharma M et al (2020) Inferring the effectiveness of government interventions against COVID-19. Science. https://doi.org/10.1126/science.abd9338
19. Zhou P et al (2020) A pneumonia outbreak associated with a new coronavirus of probable bat origin. Nature 579:270–273
20. Petrosillo N, Viceconte G, Ergonul O, Ippolito G, Petersen E (2020) COVID-19, SARS and MERS: are they closely related? Clin Microbiol Infect 26:729–734
21. Jaimes JA, Millet JK, Whittaker GR, Proteolytic Cleavage of the SARS-CoV-2 spike protein and the role of the novel S1/S2 Site. I Science 23:101212. s. https://doi.org/10.1016/j.isci.2020.101212
22. Corman VM, Muth D, Niemeyer D, Drosten C (2018) Hosts and sources of endemic human coronaviruses. Adv Virus Res 100:163–188
23. Guarner J (2020) Three emerging coronaviruses in two decades. Am J Clin Pathol 153:420–421
24. Romano M, Ruggiero A, Squeglia F, Maga G, Berisio R (2020) A structural view of SARS-CoV-2 RNA replication machinery: RNA synthesis, proof reading and final capping. Cells 9:1267
25. Kang S, Yang M, Hong Z, Zhang L, Huang Z, Chen X et al (2020) Crystal structure of SARS-CoV-2 nucleocapsid protein RNA binding domain reveals potential unique drug targeting sites. Acta Pharm Sin B 10:1228–1238
26. Zeng W, Liu G, Ma H, Zhao D, Yang Y, Liu M et al (2020) Biochemical characterization of SARS-CoV-2 nucleocapsid protein. Biochem Biophys Res Commun 527:618–623
27. Henderson R, Edwards RJ, Mansouri K, Janowska K, Stalls V, Gobeil SMC et al (2020) Controlling the SARS-CoV-2 spike glycoprotein conformation. Nat Struct Mol Biol 27:925–933
28. Huang C, Wang Y, Li X, Ren L, Zhao J, Hu Y et al (2020) Clinical features of patients infected with 2019 novel coronavirus in Wuhan, China. Lancet 395:497–506
29. Astuti I, Ysrafil (2020) Severe acute respiratory syndrome coronavirus 2 (SARS-CoV-2): an overview of viral structure and host response. Diabet Metab Syndr 14:407–412
30. Tang X, Wu C, Li X, Song Y, Yao X, Wu X et al (2020) On the origin and continuing evolution of SARS-CoV-2. Natl Sci Rev 7:1012–1023
31. Schoeman D, Fielding BC (2019) Coronavirus envelope protein: current knowledge. Virol J 16:69
32. Tai W, He L, Zhang X, Pu J, Voronin D, Jiang S et al (2020) Characterization of the receptor-binding domain (RBD) of 2019 novel coronavirus: implication for development of RBD protein as a viral attachment inhibitor and vaccine. Cell Mol Immunol 1e8
33. Chen Y, Liu Q, Guo D (2020) Emerging coronaviruses: genome structure, replication, and pathogenesis. J Med Virol
34. Walls AC, Park Y-J, Tortorici MA, Wall A, McGuire AT, Veesler D (2020) Structure, function, and antigenicity of the SARS-CoV-2 spike glycoprotein. Cell 181:281–292

35. Wrapp D, Wang N, Corbett KS, Goldsmith JA, Hsieh C-L, Abiona O et al (2020) Cryo-EM structure of the 2019-nCoV spike in the prefusion conformation. Science 367:1260–1263
36. Ortega JT, Serrano ML, Pujol FH, Rangel HR (2020) Role of changes in SARS-CoV-2 spike protein in the interaction with the human ACE2 receptor: an in-silico analysis. EXCLI J 19:410–417
37. Lan J, Ge J, Yu J, Shan S, Zhou H, Fan S et al (2020) Structure of the SARS-CoV-2 spike receptor-binding domain bound to the ACE2 receptor. Nature 581:215–220
38. Shang J, Wan Y, Luo C, Ye G, Geng Q, Auerbach A et al (2020) Cell entry mechanisms of SARS-CoV-2. Proc Natl Acad Sci USA 117:11727–11734
39. Xia S, Liu M, Wang C, Xu W, Lan Q, Feng S et al (2020) Inhibition of SARS-CoV-2 (previously 2019-nCoV) infection by a highly potent pan-coronavirus fusion inhibitor targeting its spike protein that harbors a high capacity to mediate membrane fusion. Cell Res 30:343–355
40. Zhang S, Qiao S, Yu J, Zeng J, Shan S, Tian L, Lan J, Zhang L, Wang XJNC (2021) Bat and pangolin coronavirus spike glycoprotein structures provide insights into SARS-CoV-2 evolution. Nat Commun 12:1–12
41. Bajaj A, Purohit HJ (2020) Understanding SARS-CoV-2: genetic diversity, transmission and cure in human. Indian J Microbiol 1
42. Brufsky A (2020) Distinct viral clades of SARS-CoV-2: implications for modeling of viral spread. J Med Virol
43. Islam MR et al (2020) Genome-wide analysis of SARS-CoV-2 virus strains circulating worldwide implicates heterogeneity. Sci Rep 10:1–9
44. Wang C et al (2020) The establishment of reference sequence for SARS-CoV-2 and variation analysis. J Med Virol 92:667–674
45. Yuan M, Wu NC, Zhu X, Lee C-CD, So RTY, Lv H et al (2020) A highly conserved cryptic epitope in the receptor-binding domains of SARS-CoV-2 and SARSCoV. Science (80):368, 630–633
46. Kirchdoerfer RN, Wang N, Pallesen J, Wrapp D, Turner HL, Cottrell CA et al (2018) Stabilized coronavirus spikes are resistant to conformational changes induced by receptor recognition or proteolysis. Sci Rep 8:1–11
47. Sorokina M, Teixeira JMC, Barrera-Vilarmau S, Paschke R, Papasotiriou I, Rodrigues JP et al (2020) Structural models of human ACE2 variants with SARS-CoV-2 Spike protein for structure-based drug design. Sci Data. 7:1–10
48. Clausen TM, Sandoval DR, Spliid CB, Pihl J, Perrett HR, Painter CD et al (2020) SARS-CoV-2 infection depends on cellular heparan sulfate and ACE2. Cell 183:1043–1057
49. Banner LR, Lai MM (1991) Random nature of coronavirus RNA recombination in the absence of selection pressure. Virology 185:441–445
50. Bande F, Arshad SS, Omar AR, Hair-Bejo M, Mahmuda A, Nair V (2017) Global distributions and strain diversity of avian infectious bronchitis virus: a review. Anim Health Res Rev 18(1):70–83
51. Gandon S, Day T (2008) Evidences of parasite evolution after vaccination. Vaccine. https://doi.org/10.1016/j.vaccine.2008.02.007
52. Gandon S, Mackinnon MJ, Nee S, Read AF (2001) Imperfect vaccines and the evolution of pathogen virulence. Nature 414(6865):751–756
53. García de Alcañíz J, López-Rodas V, Costas EJM (2021) Sword of damocles or choosing well. In: Population genetics sheds light into the future of the COVID-19 pandemic and SARS-CoV-2 new mutant strains
54. Xu J, Zhao S, Teng T, Absalla EA, Zhu W, Xie L et al (2020) Systematic comparison of two animal-to-human transmitted human coronaviruses: SARS-CoV-2 and SARS-CoV. Viruses 12(2):244
55. Wan Y, Shang J, Graham R, Baric RS, Li FJ (2020) Receptor recognition by the novel coronavirus from Wuhan: an analysis based on decade-long structural studies of SARS coronavirus. J Virol. https://doi.org/10.1128/JVI.00127-20
56. Singh P, Ocampo M, Lugo JE, Doti R, Faubert J, Rawat S et al (2018) Fractal and periodical biological antennas: hidden topologies in DNA, wasps and retina in the eye. In: Ray K, Pant M, Bandyopadhyay A (eds) Soft computing applications. Springer, Singapur, pp 113–130

57. Ghosh S, Chatterjee S, Roy A, Ray K, Swarnakar S, Fujita D (2015) Resonant oscillation language of a futuristic nano-machine-module: eliminating cancer cells & Alzheimer Aβ plaques. Curr Top Med Chem 15(6):534–541
58. Wiltshire MCK, Hajnal JV, Pendry JB, Edwards DJ, Stevens CJ (2003) Metamaterial endoscope for magnetic field transfer: near field imaging with magnetic wires. Opt Express 11(7):709–715
59. Rehman HA et al (2021) Comprehensive comparative genomic and microsatellite analysis of SARS, MERS, BAT-SARS, and COVID-19 coronaviruses. J Med Virol
60. Bandyopadhyay A, Sahu S, Fujita D, Wakayama Y (2010) A new approach to extract multiple distinct conformers and co-existing distinct electronic properties of a single molecule by point-contact method. Phys Chem Chem Phys 12(9):2198–2208
61. Bandyopadhyay A, Wakayama Y (2007) Origin of negative differential resistance in molecular junctions of Rose Bengal. Appl Phys Lett 90(2):023512
62. Ghosh S, Sahu S, Agrawal L, Shiga T, Bandyopadhyay A (2016) Inventing a co-axial atomic resolution patch clamp to study a single resonating protein complex and ultra-low power communication deep inside a living neuron cell. J Integr Neurosci 15(04):403–433
63. Ghosh S, Sahu S, Fujita D, Bandyopadhyay A (2014b) Design and operation of a brain like computer: a new class of frequency-fractal computing using wireless communication in a supramolecular organic, inorganic systems. Information 5:28–99
64. Saxena K, Singh P, Sahoo P, Sahu S, Ghosh S, Ray K et al (2020) Fractal, scale free electromagnetic resonance of a single brain extracted microtubule nanowire, a single tubulin protein and a single neuron. Fractal Fract. https://doi.org/10.3390/fractalfract4020011
65. Singh P, Ray K, Fujita D, Bandyopadhyay A (2019) Complete dielectric resonator model of human brain from MRI data: a journey from connectome neural branching to single protein. In: Ray K, Sharan S, Rawat S, Jain S, Srivastava S, Bandyopadhyay A (eds) Lecture notes in electrical engineering. Springer, Singapore, pp 717–733
66. Ghosh S, Dutta M, Sahu S, Fujita D, Bandyopadhyay A (2014a) Nano molecular-platform: a protocol to write energy transmission program inside a molecule for bio-inspired supramolecular engineering. Adv Func Mater 24:1364–1371
67. Ghosh S, Fujita D, Bandyopadhyay A (2015) An organic jelly made fractal logic gate with an infinite truth table. Sci Rep 5:11265
68. Reddy S, Sonkar D, Singh P, Saxena K, Singh S, Chhajed R et al. (2018) A Brain-like computer made of time crystal: could a metric of prime alone replace a user and alleviate programming forever? in Studies in computational Intelligence, 761, 1–44. Springer Nature Singapore Pvt. Ltd, ISBN: 978-981-10-8048-7; https://doi.org/10.1007/978-981-10-8049-4_1
69. Winfree A (1977) Biological rhythm research 8, 1; the geometry of biological time, 2nd edn. Springer, New York
70. Yao NY, Potter AC, Potirniche ID, Vishwanath A (2017) Discrete time crystals: rigidity, criticality, and realizations. Phys Rev Lett 118:030401–030406
71. Zhang Z, Hess PW, Kyprianidis A, Becker P, Lee A, Smith J et al (2017) Observation of a discrete time crystal. Nature 54:217–220
72. Bruinsma RF, Gelbart WM, Reguera D, Rudnick J, Zandi R et al (2003) Viral self-assembly as thermodynamic process. Phys Rev Lett 90(24):248101–248111
73. Bandyopadhyay A (2020a) Nanobrain: the making of an artificial brain from a time crystal. Taylor & Francis Inc. Imprint CRC Press Inc., Bosa Roca, United States, p 336. ISBN 10-1439875499. ISBN 13-9781439875490. https://doi.org/10.1201/9780429107771
74. Bandyopadhyay A, Ghosh S, Fujita D (2020b) Universal geometric-musical language for big data processing in an assembly of clocking resonators, JP-2017-150171, 8/2/2017: World patent, WO 2019/026983; US Patent App. 16/635,900
75. Bandyopadhyay A, Ghosh S, Fujita D (2020c) Human brain like intelligent decision-making machine; JP-2017-150173; 8/2/2017; World patent WO 2019/026984; US Patent App. 16/635,892
76. Singh P, Saxena K, Singhania A, Sahoo P, Ghosh S, Chhajed R et al (2020) A self-operating time crystal model of the human brain: can we replace entire brain hardware with a 3D fractal architecture of clocks alone? Information 11(5):238

77. Vernizzi G, Sknepnek R, de la Cruz MO (2011) Platonic and Archimedian geometries in multicomponent elastic membranes. PNAS 108:4292–4296
78. Singh P et al (2020b) A space-time-topology-prime, stTS metric for a self-operating mathematical universe uses dodecanion geometric algebra of 2–20 D complex vectors. In: Ray K, Roy KC, Toshniwal SK, Sharma H, Bandyopadhyay A (eds) Proceedings of international conference on data science and applications. Lecture notes in networks and systems, vol 148. Springer, Singapore. https://doi.org/10.1007/978-981-15-7561-7_1
79. Singh P et al (2020c) Quaternion, octonion to dodecanion manifold: stereographic projections from infinity lead to a self-operating mathematical universe. In: Singh P, Gupta RK, Ray K, Bandyopadhyay A (eds) Proceedings of international conference on trends in computational and cognitive engineering. Advances in intelligent systems and computing, vol 1169. Springer, Singapore. https://doi.org/10.1007/978-981-15-5414-8_5
80. Bandyopadhyay A, Sahu S, Fujita D (2009) Smallest artificial molecular neural-net for collective and emergent information processing. Appl Phys Lett 95(11):113702
81. Milosevic M (2013) On the nature of the evanescent wave. Appl Spectrosc 26(2):126–131
82. Nye JF (1983) Polarization effects in the diffraction of electromagnetic waves: the role of disclinations. Proc R Soc Lond A387:105–132
83. Agrawal L, Chhajed R, Ghosh S, Ghosh B, Ray K, Sahu S et al (2016) Fractal information theory (FIT) derived geometric musical language (GML) for brain inspired hypercomputing. In: Pant M, Ray K, Sharma T, Rawat S, Bandyopadhyay A (eds) Soft computing: theories and applications. advances in intelligent systems and computing. Springer, Singapore, pp 343–372
84. Aschoff J, Wever R (1981) The circadian system of man. In: Aschoff J (ed) Biological rhythms. Springer, MA, Boston
85. Gurevich Y, Shelah S (1989) Nearly linear time, vol 363. Springer, LNCS, pp 108–118
86. Pippinger N, Fischer MJ (1979) Relations among complexity measures. J ACM 269(2):361–381
87. Malinowski JR, Laval-Martin DL, Edmunds LN Jr (1985) Circadian oscillators, cell cycles, and singularities: light perturbations of the free-running rhythm of cell division in Euglena. J Comp Physiol 155B:257–276

Chapter 2
Cloaking or Invisibility is the Foundation of Highly Intelligent Bio-machinery: Why Are Computer Circuits Primitive?

2.1 Introduction

2.1.1 The Brilliant Engineering of Proteins as an Electromagnetic Engine

Knowing the basic mechanism of information processing in the biological system is an exciting current topic of research. However, the problem is that no scientist and thinker could forget the huge movement of creating the information structure for decision-making in the 1940s. The information is "bits," one among multiple choices. Information cannot be multiple choices because then we need to know who makes the choices. The observer? Who guarantees that the observer observed the reality? Since the last hundred years, proteins have been investigated as analogous to some market-available electronic devices. Since the 1930s, the electromagnetic resonance of biomaterials has been measured as transmission coefficient. It means the ratio of amplitudes of the incident and transmitted waves [1–3]. Therefore, we could crudely conclude that the materials turn transparent to ac signal or alternating current at resonance frequencies. Transparency is not cloaking. Cloaking means electromagnetic waves pass through a material and return to the initial path after passing through. So, the image at the back appears in the front, the object in the middle disappears. This is cloaking [4].

If proteins cloak in the living cell and the cellular network, it will open up a new era of biophysics. The cable-like transport of ions and molecules that was the foundation of all forms of known communications in a biological system would require a change. Especially, signal and energy transport via electromagnetic signals would change everything we know in biology. There are plenty of research works since the 1930s which have experimentally demonstrated that biomaterials exhibit sharp electromagnetic resonance peaks. However, despite several evidence of electromagnetic resonances of biomaterials, no single biological phenomenon has been attributed to electromagnetic resonance. Though biological systems function using chemicals,

P. Singh et al., *Biological Antenna to the Humanoid Bot*, Studies in Rhythm Engineering,
https://doi.org/10.1007/978-981-16-9677-0_2

interactions we know are chemicals, wireless energy exchange via electromagnetic resonance would coexist and cooperate.

2.1.2 Ten Advantages of a Cloaking Protein

Proteins have wide ranges of temporal operations; some events happen in and around protein for seconds and hours to a few picoseconds. Time taken to execute a particular event could be equivalent to a waveform with an equivalent period. So, time and frequency could be correlated in protein.

1. By applying suitable frequencies, one could turn a protein invisible. Selective visible of desired proteins could transmit signals through the desired route.
2. Particular secondary structures of a singular protein could be made invisible. So, we could activate or deactivate particular functionalities of a single protein.
3. Self-assembled proteins creating a crystal-like ordered assembly could vanish like a single object.
4. Thin water channels that attach to the protein molecule play a pivotal role in cloaking, hence without vortex shape, all em signal damps.
5. Geometric relation between resonance frequencies could hold a geometric shape. In the presence of cloaking, many geometric shapes coexist.
6. If a protein makes one vortex of field, it can create a structure that could create an entire vortex condensate by self-assembly.
7. A protein is made of many periodic vibrations or clocks. Only the differential clocks write their key definitions like phase, amplitude, and singularities in the emitted vortex selected by cloaking.
8. A protein can transport using a 3D invariant structure made of vortices. The vortices could be made of ions, molecules, electric, magnetic, mechanical, and electromagnetic fields.
9. A polymer chain follows a tear drop to ellipsoid dynamic shape change to create a protein structure. Similar morphogenesis creates periodic cloaking events with a specific energy transmission protocol.
10. Polyatomic time crystal is a new kind of time crystal that we have introduced. The information architecture of a protein, such as the 3D assembly of the clocks, cloaking, etc., decides which clocks appear or disappear.

2.1.3 Background Literature Review of Electromagnetic Interactions of Proteins

Since the conventional idea of information processing considers either electrical potential as information, wherein stored molecules, charge, even a typical conformation could be an information content. How do we interpret bit or binary reversible states is a choice of the experimentalist; how does he design an experiment? Therefore, any biological system can store and conduct the information. Research and development have been encouraged and taken the advanced shape in material science, molecular biology, chemistry, and engineering [5]. We could think that electromagnetic resonance frequency would be treated similarly, but it is not.

The interaction of the electromagnetic wave with biomaterials could have important applications in medicine. For the past few decades, the field has grown significantly. Several conventional methods and concepts have been developed. Electromagnetic waves are used in biomaterial measurements. Microwave tomography, cancer detection using ultra-wideband signals, healthcare informatics, hyperthermia of tumors are examples. Advanced research in biomaterials is the key to pushing the boundary of biomedical technology. Such studies address the evolution of the health issue due to microwave field emission. Conductivity shifting measurements inside the human body are carried out by applying the weak electric field at the radio frequency [6]. To detect breast tumors, microwave imaging techniques are carried out to track the group of magnetotactic bacteria microbots injected into the breast [7].

Information in a biological structure is complex. It is preserved in the geometry of the objects, self-replication, and the environment conditions [8] that provide a fundamental principle hidden behind the working mechanism of the living organism. Biological systems interact with the different kinds of signals available in the environment, and several studies show their robustness property [9–14]. Mathematical models are also used to describe the mechanism of signal processing in biological systems [15]. Using experimental and mathematical studies, the geometry of the biological system renders the output function. The objective of system biology is to create an accurate model of biological organs and explore the basic hidden mechanism inside them [11]. In our previous works, we reported the interference of the E. M. field with the biological structure in which fractality, resonance, and geometry are considered the key features [16–19].

We can detect different biomaterial resonances by designing their clocking model [20]. The clocking model is a combined map of the cavity and dielectric resonator. The cavity traps the signal at a particular frequency that exhibits maximum amplitude (Fig. 2.1). Beyond this frequency, we get minimum amplitude. Similarly, a dielectric resonator also vibrates with maximum amplitude when it absorbs a suitable frequency signal equal to the energy between the two vibrational states [21, 22]. Therefore, to build a clock assembly or time crystal model, we can use two types of resonance, cavity resonance, and dielectric resonance. When we pump electromagnetic signals of different frequencies into a biomaterial [23], cavity and dielectric slab trap the signals at the resonance frequency (Fig. 2.1). It produces oscillations of the absorbed signals while other signals are reflected from or transmitted to the biomaterial. So, the biomaterial is the integrated map of cavities and dielectric resonators.

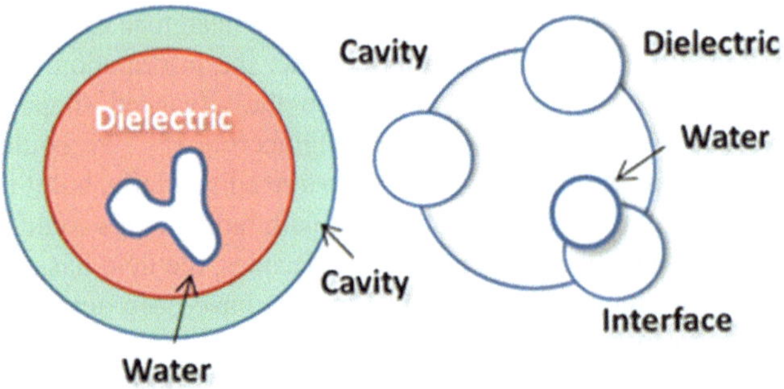

Fig. 2.1 Brain is the fusion of dielectric and cavity resonators, and each of its components contributes to building the nested time crystal model of the human brain [24]

2.1.4 Redefining Electromagnetic Resonance of Biomaterials: Why is It so Important?

We have redefined the electromagnetic resonance of biomaterials. Thus far, for the last century, we have seen that electromagnetic resonance only meant mere reflection, transmission, and absorption by the biomaterial. For the first time, we introduced additional effects in the presence of electromagnetic resonance. Initially, we started with the change in dc conductivity of a material when it is resonantly vibrating [25–27]. It means we should consider secondary responses of the materials under resonance. It could be light–matter interaction, where monochromatic polarized light falling on the resonantly oscillating matter would change angular momentum and carry surface charge density profiles. Electromagnetic resonance frequencies of the biomaterials are found in a wide frequency range (kHz–THz) in the simulation and experimental studies. The effect of the electromagnetic wave on the cell membrane, transmembrane potential, and cell cycle are detected in MHz, GHz, THz frequency range [28, 29]. Saxena et al. showed that ion density distribution on the surface of biological organisms changes during the electromagnetic wave interaction. A most important discovery was that the time domain where ionic resonance peaks are 1000 times longer than the time domain where dipole resonance peaks are normally observed. Nevertheless, the grouping of resonance frequencies follows a similar pattern for both the time domains. It means biological structures are so designed that we follow the conventional biological paradigm where molecules and ions do everything chemically. Alternatively, we follow our radically orthogonal paradigm electromagnetic packet of energy as a vortex or ring of fields. In both cases, the relation between frequencies at which the vortex generates remains similarly arranged.

2.1.5 *Electromagnetic Resonance of DNA: How Does It Differ from Proteins*

Electromagnetic resonance affects the biological system when a system undergoes biological processes such as structural growth [30] from the genetic codes. Millimeter-wave alters the active carriers along the cell membrane, and it has also been reported that terahertz radiation can interface with DNA and may lead to genomic instability. Nature employs a fraction of the electromagnetic spectrum of any cell or living organism, although they are associated with the electric fields and flux of various charged atoms. The static and alternative electric fields lie in the kilohertz frequency range. A theoretical assumption suggests that living cells oscillate in the MHz–THz frequency range [31].

Electromagnetic fields of extremely low frequency (EM-ELF) directly affect intracellular DNA [32]. EM-ELF leads to the effects on health and could modulate the functional mechanism of the peripheral nervous system (PNS) and central nervous system (CNS) [33]. DNA structure follows the properties of the fractal antenna and can interact with electromagnetic fields over a wide frequency range. DNA emits radiation like a helical antenna. Insulators like biomaterials do not have the highly reflective metallic boundary so they cannot generate the high-quality standing wave [16]. The electromagnetic resonances of biological molecules like DNA, RNA, and proteins may be important for the function of macromolecules and may be useful in the development of devices like the molecular computer. The results were not found in the order of principle [34]. Experimental realization of the macromolecules resonances is expensive and time-consuming, so computational method like the Resonant Recognition Model (RRM) is used to predict the electromagnetic resonance of tubulin and microtubule that have been experimentally confirmed [35].

2.2 Ten Unique Features of Novel Electromagnetic Resonance

1. In the majority of the biological structures, the electric and the magnetic field split on the surface of the material at resonance frequencies. In the structural part where the electromagnetic energy rotates in a loop, we find the dominance of the magnetic field, and where the energy flow is linear, the electric field dominates.
2. Both electric and magnetic parts of the resonant energy absorbed by the material rotate in a loop on the surface of the resonating material. If we change the phase of the input signal, we can see the motion of electric and magnetic fields following the loop path.
3. The dimension of the biomaterial plays a crucial role in the interference of the flowing magnetic and electric fields at resonance. By varying the dimensions like length, pitch, and diameter, we get a phase structure with twelve holes [1].

At different dimensions, lengths, curvature, and diameter, some holes close and some open.

4. The resonantly vibrating biomaterial hinders the flow of direct and alternating current throughout the biomaterial's surface. The electronic energy density becomes such that the same material could respond or interact to incident energy differently. We could suggest that by tuning frequencies, we could modulate the geometry of the surface for various applications like metamaterials, cloaking, superlens, etc.
5. The stored charge on the resonantly vibrating biomaterial could linearly modulate the flux generated when electromagnetic energy falls on its surface. The resonantly vibrating biomaterial absorbs a part of the incident energy. A part of that is added to the vortex produced as a phase, which eventually contributes to the energy of the vortices. The field of vortices accounts for the flux. Thus, the stored charge, on the surface and the flux, is produced to develop a linear or nonlinear relationship.
6. Biomaterials, because of their extremely large resonance band, spanned over 10^{12} orders of time scale, starting from THz to Hz. Biomaterials respond uniquely to electromagnetic or electrical, or magnetic noise bursts. Any form of noise is a random uncorrelated composition of signals at various frequencies. For that reason, we found them harvesting the thermal and various other forms of noise.
7. Most biomaterials are insulators. They act as capacitors and store charges. In the textbooks, we are taught that proteins transmit ions. Therefore, it is a conducting material. However, note that controlling these ion flows externally by applying an electric field or any other form of forces makes it impossible to modulate the rate of ion flow. Thus, only to protect biological phenomenon, nature has designed protein such that its ion transport mechanism does not get affected by external fields or signals. Imagine, if they do, ion channels could modulate ion transport, which would mean the death of a cell. Consequently, the proteins turn critically ferroelectric. We could observe perfectly square current–voltage characteristics [25].
8. Most biomaterials are piezoelectric since ions store differently in different secondary structures of proteins or adopt different potentials in different parts of the same molecular system. As a result, a biomaterial could be considered an integrated structure of cavities, wherein each cavity has a distinct electric potential. When we mechanically press a protein, we force a cluster of secondary structures with distinct potentials to exchange electrons and redistribute charge density. Conversely, if electric bias is applied, these local secondary structures exchange charge, and secondary structures vibrate to adjust modified conformation to reach a new equilibrium.
9. Since proteins have multiple clusters of secondary structures. These isolated clusters act as superlens since electric and magnetic fields occupy defined spatial pathways. The photons entering these structures change phase as a function of the space-time matrix but return to the same phase after passing through the material. As a result, photons passing through the material acquire

initial spatial coordinates. Therefore, the mechanism of quantum cloaking that we reported for proteins earlier [1, 4] is not similar to the one we often find in the literature. In contrast, we showed earlier that tubulin protein has eight cavities. Each cavity with a group of secondary structures acts as a distinct resonating center as imaged by quantum tunneling [35]. All eight could resonate at different frequencies. It means proteins do electromagnetic lensing at different frequencies. One could imagine a protein-like combination of transparent glass cubes, which could turn opaque suitably. Therefore, the system could hold a space–time metric of photon and return on passing through.

10. Symmetry of primes in grouping the resonance frequencies have been the most profound observation that we made. We could see explicitly that fifteen primes are widely used in grouping the resonance frequencies. It means if we monitor the ratio of the resonance frequencies and find selective frequencies that change phase synchronously. Based on the phase loop, the frequencies are grouped. Say, three resonance frequencies form a group. They change phase, say 30–35° together synchronously. Now, if the other two groups are shifting in phase 70–75° (e.g. 11 peaks) and the third group shifting in phase 40–45° (e.g. seven peaks), where there is an exact five-degree variation in phase, this can lead to three groups which form a hierarchical group. Biomaterials tend to grow within and above. Then biosystems prefer 8 and 12. However, for grouping phases, primes are widely used.

2.3 Biomaterials and Proteins as Metamaterials

2.3.1 The Early History of a Metamaterial

In the 1900s, there was the courage to design artificial materials for the manipulation of electromagnetic waves. In the 1940s, American industrial research and development created a material with the same characteristics as metamaterials. In the 1967s, negative refractive index material was first described as having the ability to transmit light [36]. Simultaneously, the phase velocity could be drawn antiparallel to the Poynting vector, which is not normally the same as wave propagation in natural materials [37]. Such materials allow the EM wave to convey the energy against the group velocity. Negative permittivity could be generated by aligning the wire along the direction of wave propagation [36]. Some natural materials like ferroelectrics show negative permittivity, but the challenge was about generating the negative permeability. Some periodic arrays of rings or wires can generate the negative refractive index, and split ring resonators can produce the negative permeability. After that, the experimental realization of functional electromagnetic metamaterials was demonstrated. A negative refractive index metamaterial could be realized using artificial lumped-element loaded transmission lines in microstrip technology. In 2007, several groups [38] experimented on the negative refractive index-based metamaterials. In 2006, for the first time, an invisibility cloak was realized at microwave frequency.

Fig. 2.2 Microtubule: alpha (gray color) and beta-tubulin (blue color)

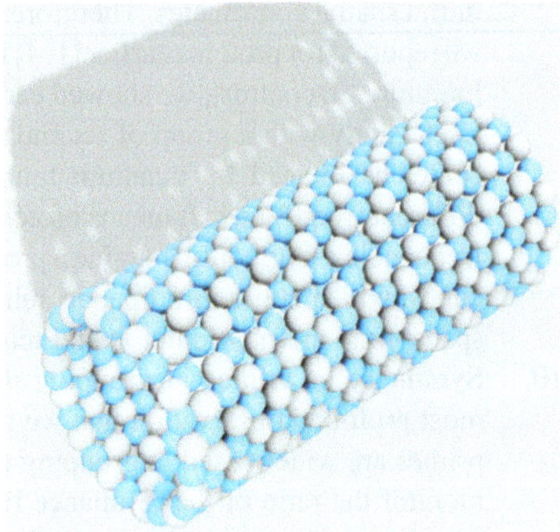

Nowadays, metamaterials have emerged as the most popular topic in the scientific community due to their unusual electromagnetic properties, for example, microtubules (Fig. 2.2). For the last few decades, a huge effort has been made to create the functional device based on metamaterial for many applications in electromagnetics, solid-state physics, material science, nanoscience and semiconductor physics, microwave and antenna engineering, semiconductor engineering, etc. [39]. The metamaterial is engineered to have properties that do not occur in natural materials [40], and they are designed by assembly of several elements. The materials are arranged in the array form so that the dimension of the device is less than the wavelength of light so that they influence the light. Metamaterials do not show their properties due to the base materials but rather the new material's design, shape, orientation, and size. They provide unusual properties by manipulating the electromagnetic wave by enhancing, bending, blocking, and absorbing the electric and magnetic vectors. Such materials affect the electromagnetic wave more than conventional materials [41].

Furthermore, it shows the negative refractive index for a particular light wavelength, known as negative-index metamaterial [42]. To realize metamaterial applications, the energy loss must be reduced and material must be scaled up to 3D in the isotropic form [43]. For example, a superlens is a 3D or 2D device made from metamaterials with negative refraction properties that achieve the resolution beyond their diffraction limit. To achieve such behavior, double negative materials are used to produce negative phase velocity. The diffraction limit is inbuilt in the lenses [44, 45]. The metamaterial is the basis for a practical cloaking device [46, 47].

2.3.2 The Necessity of Water Layer

Water is a resource of life, and sometimes it causes a mystery of its solvation properties. It is the medium of dissipated energy and mediates the interactions in the biological systems, for example, DNA. Chiral water forms the superstructure around a biomolecule. In that case, water molecules that follow the helical structure of DNA that is also chiral. If we want to understand the reactivity of water, it is not only to understand water. We have to understand water's reactivity, behavior, and interaction around biological materials like DNA and protein. Water plays a key role in understanding the DNA structure and functions. The hydration shell of water is a current topic of research. The detailed behavior of water structure in the major and minor grooves of DNA could be understood by Molecular Dynamic Simulation, MDS.

The behavior of biomolecules such as DNA and microtubule redefines when the different-shaped water layers are wrapped around them. In the presence of the water layer, biomaterials turn into the metamaterial at a particular frequency by changing the signal transmission nature in itself. Metamaterial controls the wave propagation in a specific frequency range of the electromagnetic spectrum that has the potential to make the object invisible from our view. We also carried out a detailed analysis of the electric and magnetic field separation in the biomaterials.

Cloaking materials are metamaterials with a property to be invisible in front of our eyes. Thus far, no biomaterial is found to disappear, revealing what is at its back. Recently, a helical water channel was found in DNA intertwined with its double helix. We theoretically observed that a helical water channel has an incredible property to be invisible and turn material around it invisible. When we found that it turns DNA invisible, we took a quantum tunneling image using a specially developed scanning tunneling microscope at different electromagnetic frequencies. We observe that DNA slowly disappears as the applied frequency reaches the theoretically predicted cloaking frequency, and visibility returns gradually. The disappearance of quantum tunneling, revealing the surface beneath, is shocking because it demands changing the concept of wave function interaction with electromagnetic waves; the simple addition of electromagnetic potential with the quantum well potential is oversimplified. A material could bend quantum wave function like a classical wave function route. An electromagnetic wave has an electric and a magnetic vector. They bend distinctly while passing through a material following a refractive index, primarily electric susceptibility and magnetic permeability.

2.4 Time Crystal Composites: Classical and Quantum Metamaterial

The light said, go beyond the glass, and then there was metamaterial, we achieved the ability to build a composite of lenses to synthesis arbitrarily complex waveforms. Then, the shape of a lens, be it convex or concave, does not matter. It is the composition of matter that shapes the wavefront of light in the desired way. In a classical metamaterial, when a large waveform splits into many local waveforms due to local curvatures, only one singular large waveform is generated. That singular waveform is the modulation of all constituent local wavefronts produced at distinct frequencies; one could see those signals using an FFT measuring device. In the quantum metamaterial, the tunneled signal is a superposition of locally produced several independent waveforms connected by multiple coexisting phase relations between them. The original information is destroyed if isolated from others because it does not just lose relative phase relations with the neighboring clocks. However, a composition of clocks builds new clocks, which disappear. It means, simply by shining electromagnetic waves, we get a composition of clocks that are interconnected as if many spheres are put together into one singular structure. Great circles of the spheres are clocks. System points run along the perimeter, typical waveforms all connected by specific phase relations are produced. Modulated wave that forms in the classical domain as interference of multiple waveforms is replaced by an architecture of clocks in the quantum domain. Modulated wave in classical interaction with the metamaterial contains one singular system point. However, in quantum, there are plenty of system points that run in parallel. Since this architecture resembles the unit cell of a time crystal, we call it a time crystal.

2.4.1 Anisotropy of Biomaterials Splits E and B, Forms an Evanescent Wave

The lens helps us to see small structures in detail. Superlens does not amplify. The resolution of imaging offered by superlens does not depend on the wavelength of electromagnetic waves, but the material's geometry encodes intricate local structural information in the evanescent wave [45]. The evanescent wave is in the opposite phase with the electromagnetic wave from which it is born. This is normal for reflected waves. As the signal phase reverses, while canalizing energy from the incident electromagnetic wave and the emitted evanescent wave, the noise in the input signal is naturally eliminated in the evanescent wave. The evanescent wave is pure, noise-free. It breaks the diffraction limit of light. Once a near field image is resolved using an evanescent wave, one has to project it far, which is called hyperlens. Microtubule acts as an antenna and superlens at a time, it combines superlens and hyperlens. In 2014, the maximum resolution achieved for superlens was 35 nm [48], however, 50–70 nm resolution imaging is frequent now [49].

Above diffraction limit (>200 nm), the structures reflect light; below, inhomogeneous structures like holes or tiny protrusions build anisotropy, i.e., wave vectors in different directions flow differently. Electron density distribution on the surface absorbs the electric vector, while an electron density flow in a loop interacts with the magnetic vector. In a highly curved material, within a very small area, somewhere, the electric vector is absorbed more, and at some locations, the magnetic vector is absorbed more. As a result, an electromagnetic wave, whose electric and magnetic parts are fundamentally coupled, generates an undefined field region or singularity where the complementary signal parts burst out to conserve energy, namely, evanescent wave. Thus, anisotropy splits an electromagnetic wave's electric and magnetic parts [50]; they amplify evanescent waves by increasing anisotropy). Consequently, electric and magnetic fields distribute unevenly around the local domain, and shined electromagnetic waves create local oscillations, namely, evanescent waves. In the evanescent wave, electric and magnetic fields are largely decoupled [51]. Thus, the true nature of an evanescent wave is not a common electromagnetic wave that we know. It could be a pure electric field, purely magnetic, or even a mixture of the two. This is why when we build an electric or magnetic vortex, there is a high probability that we get a structure of field distribution made of purely electric or purely magnetic.

2.4.1.1 Asymmetric Treatment of an Electromagnetic Wave

The magnetic study of biomaterials is currently limited to applying a high magnetic field forcing the spins to align. However, even without a high magnetic field (~10^4 T) [52], heartbeat, breathing naturally generate the waves of low magnetic fields (10^{-10} T). It is a quest for long to isolate two fields of an electromagnetic wave. Isolating the magnetic part is not the realization of a monopole. A wavelike magnetic field flow is mostly generated in two ways: by asymmetrically treating an electromagnetic field or by a periodic flow of ions. The topology of the structures could regulate the beating of charges and generate a wavelike flow of magnetic field [53–55]. In the optical vortex studies, by manipulating the surface topology, electric and magnetic parts are interfered separately [56]. However, all studies are purely optical. If isolated electric and magnetic parts activate the separate regions of a biomaterial, initiate a typical dynamic flow of fields, it would open the route to wirelessly regulate a biological phenomenon [57]. While the quest for a pure magnetic wave continues as a magnetic vortex around a solid-state defect site [58], here the studied biomaterials are non-magnetic. If the electric vector lies in a plane with the surface, it interacts more with the surface electrons than the magnetic field. A surface can treat two identical vectors very differently depending on the relative wave vector orientation. If the electric vector is nullified, the electromagnetic wave changes its electric nature to a magnetic one. Isolating the magnetic part of an electromagnetic wave is the key to a global positioning sensor.

2.4.1.2 The Electric Field Prefers a Closed Loop, the Magnetic Field Prefers a Spiral In/Out

If we have a hollow sphere, outside, the dipolar fields due to electric and magnetic parts are identical. The electric moments are changed by the magnetic moments in their mathematical expression. However, inside, the electric and magnetic moments reverse their sign. In a dielectric resonator antenna ($\lambda/\sqrt{\epsilon}$), the piecewise linear distribution of elementary dipoles often varies the potential [59] with LogQ, not Q^2/r(Q = charge). Here, in a standalone non-conducting material, the reflecting boundary is not present. So, a closed-loop flows the energy of a static electric or magnetic field, forming a virtual feedback loop. That ensures a periodic oscillation, but with a difference. In the above formulation, the electric moment affects 1/3, but magnetic moments affect 2/3 as the values change in a topological space. A 2/3rd variation leads to a spiral behavior. We investigate the possibility of a spiral flow below.

If we consider any spiral or fractal path to be a sum of a repeated local symmetry, e.g., an array of disks, the lensing of electromagnetic wave (just like a lens focuses the light beam) [60] is possible. Lensing leads to the concentration of magnetic flux in a particular direction. An array of dielectric resonators act as a stack of disks. Then, the charge does not vary as LogQ; rather, we get $e(z) = e^{+}e^{ikz} + e^{-}e^{-ikz}$, $k = \omega/e\sqrt{\mu\varepsilon}$. There have been extensive studies on the LogQ distribution of charge, but the studies on the spiral distribution of charge are scarce. It is useful because when an electromagnetic wave passes through such a media, several waveforms pack within a very small angular width, the phenomenon is triggered by an array of resonators (here, our elementary resonators are microtubule), which even could lead to a phase discontinuity or singularity [61]. Then, we find that a part of the transverse component contributes to the longitudinal component. In other words, $B_z = \frac{i}{k_c}\left(\frac{\partial E_x}{\partial y} - \frac{\partial E_y}{\partial x}\right) = \frac{i}{k_c}e(z)\cdot\nabla_t \times \overrightarrow{E_t}$. This is a topological factor that adds to $Pi \neq Mi$, when$r < R$ factor described above. Thus, not just that charge distribution regulates electric field, it treats the magnetic field distribution differently. While the electric field prefers closed loops along the transverse, the magnetic field prefers a longitudinal spatial distribution, i.e., radial in/out, which is a key for effective lensing. At resonance, the oscillating field is replaced by a static field, say H_0, then $H_{eff} = k(H_0 - \omega/\gamma)$, at resonance $H_0 = \omega/\gamma$ and the angle of rotation or phase $\theta = \gamma H_1 t$. We would observe a similar response in the electromagnetic field simulations below. We would see that an electric field prefers a closed loop, and a magnetic field tries to move spirally, thus splitting them. However, if there is a lack of clocking path for the feedback loop to activate, or the energy reflecting path is modified topologically, we would see a complex distribution as shown in Fig. 2.3c.

2.4.2 *Rebuilding the Microtubule Assemblies for Electromagnetic Resonance Study: The Difference with Conventional Study*

Now we concentrate on the microtubule assemblies that work like holding the codes for regeneration of cells by cell division. We have taken only five structures for the biophysical studies. The theoretically built microtubules are assembled in CST to create the following components: axoneme of three gall midge fly, axoneme of Sciara coprophila, centriole, and cilia and flagella. Inside a cilium and a flagellum, the microtubules are arranged in a characteristic pattern known as the $9 \times 2 + 2$, called axoneme (Fig. 2.3a). Using the available detailed structures of a centriole and four

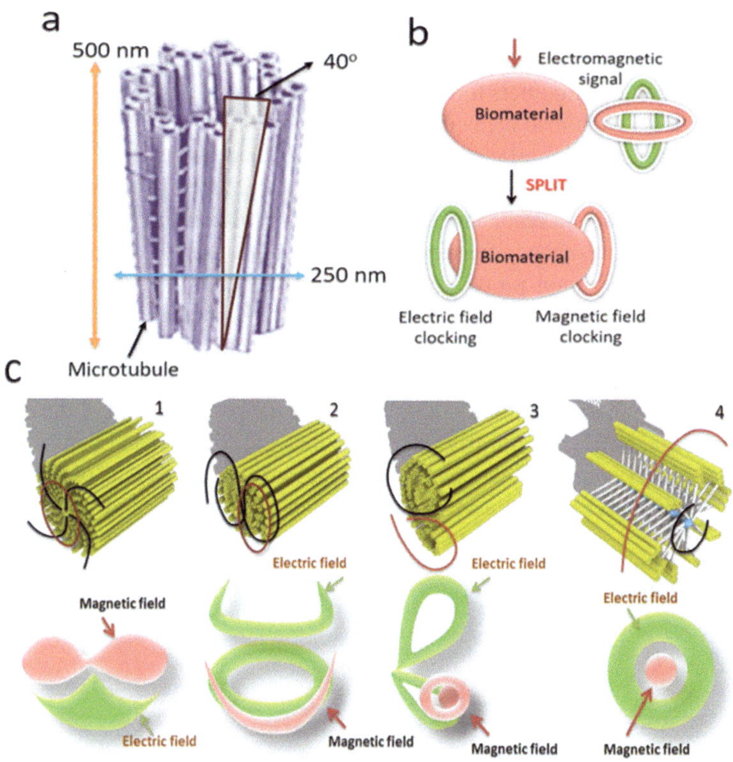

Fig. 2.3 **a** Schematic of a centriole. Out of its 27 microtubules, one is zoomed here. **b**. Splitting and periodic oscillation of electric and magnetic field as a biomaterial is pumped an electromagnetic signal of a suitable frequency. **c**. Five columns for five microtubule assemblies. The first row exhibits four realistic structures: sperm axoneme; 1. Centriole, 2. Sciara Coprophila, 3. Midge fly (I), and 4. Midge fly (III). The red lines indicate the magnetic field lines (periodic oscillation path), while the black/blue lines depict the electric field lines. Detailed field line patterns of the electric and magnetic of a bio-structure are depicted in the lower half of each panel

sperm axonemes, we have built accurate artificial structures of the biomaterials in the electromagnetic simulation software CST [62–66]. When a pair of centrioles make a centrosome, a single centriole triggers a doublet of microtubules (not triplet) growing cilium, often recognized as cell antenna [67]. However, how long would it grow is a mystery, though we know the mechanism [68]. All five structures are primarily made of the microtubule, a 25 nm wide microfilament found in the cytoskeleton of eukaryotes. While a centriole is made of a triplet of microtubule nanowires [69], the axonemes are similar but made of a doublet of microtubules.

We use nano-sized cylinder, sphere, and elementary curved 3D geometric shapes as dielectric and cavity resonators to recreate the biological structures, brick by brick, from 2 to 500 nm. In conventional dielectric resonator studies, a single geometric shape is used to mimic a giant biomaterial shape. Here, sub-nanometer alfa-helices are built first to make a single tubulin protein, and then several such proteins are used to build the microtubule. Nine such triplets of microtubules are arranged to build a centriole. A pair of SAS-6 proteins form a dimer, nine such dimers form an oligomer, nine such oligomers form a spoke of a wheel using Bld12p protein. SAS-6, Bld12p proteins were also created from scratch like tubulin protein. The dielectric constant is a function of resonance frequency, i.e., the geometry of elements used to play a vital role along with the material used. Hence, the simulated results deliver only a few fundamental resonance frequencies. Our bottom-up integration delivers a large number of resonance frequencies and a composition of dielectric constants. We kept the polarization properties of the microtubule intact, as observed in the experiments, since adding polarity to the cells is a fundamental feature of biological systems [70]. The microtubules are tilted 40° in a centriole. We included this feature. Finally, a centriole has an underlying polar structure proximal, wherein the position of the negative polarity end of a microtubule is at the proximal end of the centriole. Another new centriole is built from that proximal end. Thus, dual open cylinders at the two ends carry out the polar positioning of spindles. We took care of polarization, especially while creating the assembly of centrioles (Figs. 2.3c and 2.4).

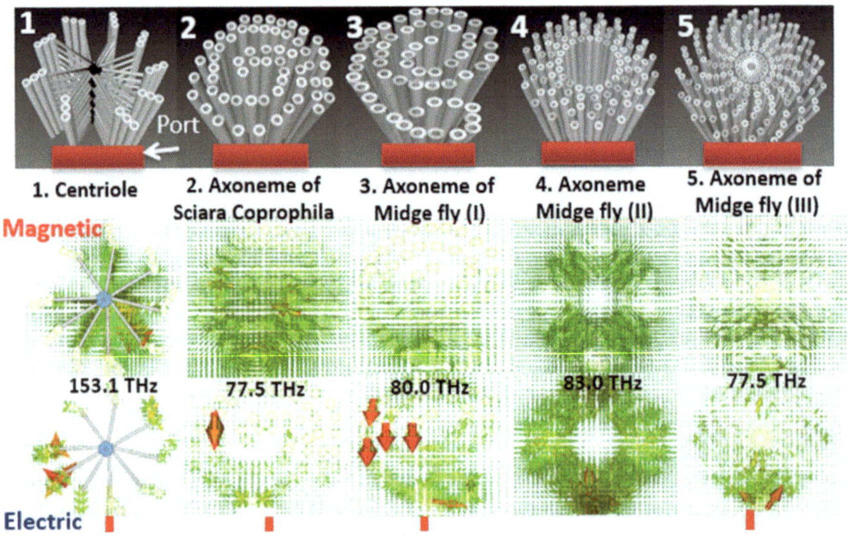

Fig. 2.4 In the top panel, device structures of Centriol, Sperm axoneme of Sciara Corprophila, and Midge fly (I), (II), (III) have been shown in panels (1), (2), (3), (4), and (5), respectively. Individually, transverse mode is set up by putting the energy supply source from the side. In the bottom panel, magnetic (upper panel) and electric (lower panel) fields distributions along the detailed structures are reported at resonance frequencies—153.7, 77.5, 80.0, 83.0, 77.5 THz

2.5 The Evolution of Microtubule Assembly into a Primitive Structure to Centriole

2.5.1 *Theoretical Simulation Protocol: Microtubule Assemblies as a Wireless Coordinate System*

We simulated the electromagnetic resonance spectrum, electric and magnetic field distribution at resonance frequencies for all the five assemblies by shifting the source (called a port) that pumps energy and the sink at various parts of the microtubule assemblies. By shifting the location of ports, we optimized for the maximum electric and magnetic resonant response. It was essential to underpin whether the clocking direction of fields, i.e., the direction of field intensity change as we change the phase from 0° to 360° is real, not a simulation artifact caused by selective placement of the measuring probe in the simulator. Only the most interesting port compositions are reported here in Figs. 2.3c and 2.4. Key features of four assemblies are: (i) a pair of sheets of microtubule spirally rotating outwards, rotating clockwise (Sciara Coprophila); (ii) a single sheet of microtubule rotating anti-clockwise and a separate small part of the same sheet rotating clockwise. It is a fusion of two rotations in opposite directions in a single structure but in two different parts of the structure (midge fly I); (iii) a complete fusion of clockwise and anti-clockwise spirals (midge fly II); and (iv) a singular clockwise spiral (midge fly III). Centriole has a spiral embedded in it. Therefore, our study explores both clockwise and anti-clockwise spiral features in the microtubule assemblies. All these data and explanations are detailed online. Both centriole and axoneme transmit signals in the THz frequency range, near the infrared region. This is consistent with the previous NIR resonance studies [71, 72]. We have confirmed both theoretically and experimentally that the electromagnetic clocks in biomaterials could run simply by available thermal noise kT. We have especially looked into the 5–6 THz domain because at room temperature ~300 K, such thermal noises are freely available in nature as an abundant source. We studied the 3D distribution of fields and the phase modulation in detail for those peaks, which provide the most significant phase-modulated energy transmission across the structure and neighboring systems in the THz frequency domain.

2.5.2 *Five Microtubule Assemblies*

We have comparatively studied five fractal microtubule assemblies. The centriole, sperm axonemes of Sciara coprophila, and three gall midge flies split an electromagnetic field's electric and magnetic parts at the specific resonance frequencies and rotate the two fields locally at two different sites. At resonance, the magnetic field prefers a spiral path that is open, while the electric field prefers a closed path, hence an electromagnetic field split. The fractal design sets an open/close path for all five assemblies to get the desired split and form infrared clocks. The spontaneous formation of non-chemical infrared clocks was tested in real 3D cellular matrices, neurons, HeLa cells, and finally in three cases where there is no centriole. It sets the energy traps far outside the physical boundary of a component with extreme precision. Earlier, a bio-rhythm was meant only for chemical rhythms or clocking chemical objects. Now, electromagnetic rhythms enable sensing the whole at a local region and from there editing the neighbor's coordinates and orientation. If we edit one with another—say, a THz wireless clock edits the circadian clock, then not just designing new molecular infrared sensors/antenna as a drug, but one could use wireless fields directly for medical treatment in the future.

2.5.3 A Generic Feature of Resonance for the Five Microtubule Assemblies

Figure 2.4 has five columns; each represents a microtubule assembly. From top to bottom, we present the raw data of 3D electric and magnetic field distributions at resonance frequencies noted in the plot. Here, for each structure, the 3D raw data of the field distribution is provided. These five structures explicitly prove the hypothesis proposed above that the geometry of biomaterials enables splitting the electromagnetic wave into magnetic and electric parts. The schematic in the lowest raw reveals the split protocol adopted by a typical microtubule assembly. For the axoneme of Midge fly III, we see that the magnetic field distributes in the structure in the form of a pair of teardrops facing each other while the electric field tries to form a closed loop. In the axoneme of Sciara coprophila, the electric field forms a closed loop, and another loop is incomplete, but the trend is visible. In contrast, the magnetic field is spiraling like an "S" shape. Structure 3 is the axoneme of midge fly I; we find that the electric field has formed a pair of teardrop loops, not a solid oscillating spiral of a teardrop. The electric field forms a closed circular loop for the centriole, while the magnetic field spirals out from the center and spirals in from the external boundary.

When we pump electromagnetic energy into a biomaterial at resonant frequencies, electric and magnetic fields are distributed in two different parts. At resonant frequencies, the electric vectors rotate on one part of the biomaterial, and the magnetic vectors rotate in other parts. So both energies are coupled but have different rotations. We found this finding in various microtubule assemblies, found in the sperm of midge flies, shown in panel c, upper part. The bottom panel is about the schematic of the rotation of electric and magnetic fields on the structures. In a centriole, the magnetic field rotates at the center of the centriole, and the electric field rotates out of the structure when we pump the electromagnetic energy in the transverse direction of the structure. In contrast, centriole preserves the same rotation of em field when we pump em energy at the longitudinal direction. Magnetic and electric fields appear as dumble and cone shapes along with the Sciara Coprophila structure. In Midge fly,

I and II, electric and magnetic fields rotate and create the loops and overlap. The combined resonance spectrum of microtubule assemblies is detected, and we have seen the resonance peaks in the THz frequency range. Of course, for biomaterials at the nanoscale, the resonance frequency would be in the THz domain. Such anisotropy nature of the em field was seen only at a resonance frequency.

Figure 2.5a, b left panels to summarize the resonance bands for five microtubule assemblies when the port is on the top and at the sides, respectively. Midge fly II and Midge fly III has a strong polar radiance (Fig. 2.4). It means they select a particular direction and pump out the resonance energy. While the axonemes of Sciara coprophila and midge fly I radiate a very low power in the similar 70–80 THz domain. Centriole's radiation is remarkable because its resonance frequency is in the blue region of visible light. However, the most important feature is that it radiates in all three directions. We have plotted the change in the field density distribution of the electric and magnetic part at a particular electromagnetic resonance frequency in Fig. 2.5a (port at top) and Fig. 2.5b (port at bottom). The field distributions for a centriole are plotted at a gap of 20° phase difference from 0° to 360°. We observed

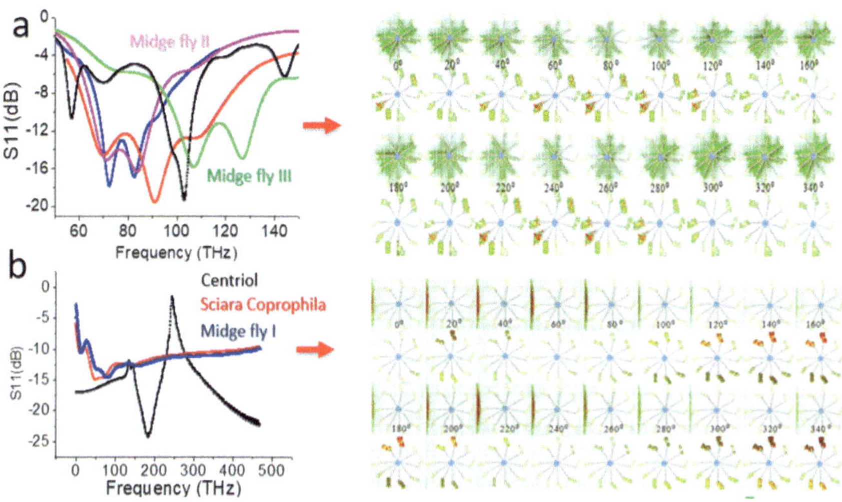

Fig. 2.5 The resonance spectra S11 (reflectance) for all five structures are shown in panel a and panel b where the energy supply ports are kept below the vertical column structure for panel b and one side of the column for panel **a** Both panels use the same color codes as noted by the color of the texts of materials names. The spatial energy distribution at the resonance peaks as a function of phase for centriole (153.1 THz, black) are plotted for both panels a and panel b to their right. The sperm axoneme of Sciara Coprophila (77.5 THz, red), and midge fly (I, II, and III) (80 THz (green III); 83 THz (pink II) and 77.5 THz, purple I). M = magnetic, E = Electric field, the corresponding phase is noted below each plot, the arrow denotes the highest energy density point. Simulation detail—Used software = CST, Used solver = Time domain solver, Boundary condition = open space, simulation frequency range = 50–150 THz, Port dimension = 7.9 × 30 nm^2, Magnetic and electric field distribution scale at resonant frequency = 153.1 THz. **b** For longitudinal mode, the resonance peak for centriole is 185.49 THz (blue). The right panel shows electric and magnetic field distribution, the simulation frequency range = 0–500 THz, port dimension = 600 × 600 nm^2

that the magnetic clocking is absent in a centriole if the applied electromagnetic field is longitudinal (Fig. 2.5a). If applied perpendicularly, the clock returns (Fig. 2.5b). Thus, the direction of energy input is the key.

2.5.4 Unique Periodic Motion of Electromagnetic Field at Resonance

These localized evanescent ripples hold key information about the tiny shapes but decay exponentially, intricate surface details are lost. In materials with a negative refractive index, the energy vector is perpendicular to the direction of growth of an evanescent wave, and phase velocity is opposite to the energy vector [73]. As a result, all discretely localized evanescent waves spatially couple, meticulous surface details are preserved. At the same time, no interaction could decay evanescent waves along the direction of growth. So, in principle, if the lens is made of a negative refractive index material, the propagating wave above diffraction limit (>200 nm) and evanescent wave below the wavelength of light (<200 nm) provide a complete picture of the surface. A negative refractive index material converges to a divergent wave. It is much more than parallelizing rays that a convex lens does.

When a biomaterial is triggered by pumping electromagnetic energy at a resonance frequency, the energy propagation across the biomaterial's surface depends on the direction of the energy source applied along with it. We have performed electromagnetic resonance studies with artificial microtubule nanowires with different tubulin (alpha and beta-tubulin) arrangements, such as the five distinct types shown in panels 1–5 of Fig. 2.6. The electric and magnetic field distributions complement the spatial distribution when the source is approximately in the longitudinal direction compared to each tubulin arrangement. They circulate in two distinct regions of the microtubule (see column 1–Center to Port). When the energy source is transverse to a microtubule length, either both E and M energy distributions are symmetric, or both the field distributions are switched on and off in a particular phase period (see second column-port from the side). In short, the charges on the surface of microtubule wire are arranged in a particular pattern at their resonance frequency, and this pattern shows different energy distributions for both locations of the energy sources.

2.5.5 The Evanescent Wave Measurement

Electromagnetic resonance of the proteins and their complexes are being measured since the 1930s. Microtubule's resonance does survive inside a neuron [25, 26, 35, 74], and it regulates the neuron firing [75]. Therefore, the key component of a centriole and an axoneme in the living system is itself an electromagnetic resonator. 3D cell–matrix study argues that nature uses in structuring the core of a neuron or a sperm

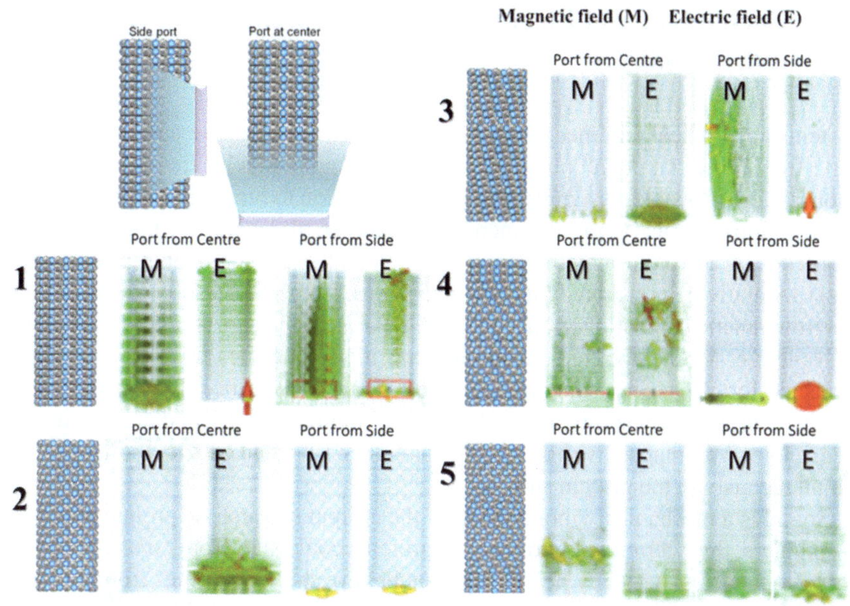

Fig. 2.6 EM energy distributions are seen on the microtubule surface. The microtubules are stimulated by pumping the energy in longitudinal and transverse mode. Here, we have carried out five types of distinct arrangements of alpha and beta-tubulin. Spatial energy distribution shows the splitting of electromagnetic energy for all types of lattice variations

cell. A microtubule assembly is not just a skeleton; they are designed to produce clocking or periodic oscillations. Possibly, this is why it has been reported that a centrosome is an infrared sensor [71], and the microtubule is called a nerve of the cell [72]. Global positioning by centrioles has a molecular origin, which requires a more rigorous investigation in the future.

Measuring the existence of an evanescent wave is easy. We have to learn the reason for the formation of an evanescent wave. We need a strong anisotropic system. For example, a vortex or spiral is a highly anisotropic system. Therefore, the transverse (perpendicular) components of the EM field which radiate the material, i.e., the wavevector components kx and ky, are decoupled from the longitudinal component kz [51]. This isolation also causes paraxiality. Suppose metallic plasma is splitting the electromagnetic energy flow into two distinct identities of the electric and magnetic field, generating the evanescent wave. In that case, we should change the width of the metallic plasma layer [50]. The amplified transmission should change with the width of the plasma layer. However, if the reason for electromagnetic field splitting is the dielectric geometry, then tuning the resonance frequencies would show variation in the transmission coefficient. Thus, the transmission coefficient across the material is a marker for the estimation of the evanescent wave.

2.6 Mechanical and Electromagnetic Resonance in Proteins and Complexes

Proteins absorb heat and mechanically vibrate between terahertz and several millihertz frequencies taking all possible values. Since no particular frequency or relaxation is more probable, assigning mechanical vibration to fundamental protein properties is impractical. A rapidly increasing number of reports on the mechanical properties of a protein molecule miss fundamental knowledge about proteins. We experimentally demonstrated that any mechanical oscillation also carries with it an electromagnetic oscillation. Both cannot be separated; the current view is half of the total picture. Most importantly, there are specific frequencies for a typical protein molecule for electrical resonance, unlike mechanical resonance. It means a protein and its complexes transmit electrical signals if and only if pumped at particular ac frequencies. We imaged live how strongly associated mechanical vibrations originate simultaneously in the protein structure.

Sound is a mechanical vibration, and if the current research on protein dynamics is complete, this vibrational frequency of proteins could start from hundreds of terahertz frequencies, basically heat. Biologists believe that heat comes from available energy or thermal fluctuations. However, they ignore that all thermal signals are electromagnetic. Even though the pure mechanical form of oscillations in proteins is widely accepted, how an electromagnetic signature of heat disappears has been a concern. Eventually, it became a pure mechanical signal that was never answered, and even such questions were hardly asked. Another fundamental concern that is always there, in the case of a single small molecule, very small relative change in position of a single atom, even fractionally less than a nanometer, the electronic potential distribution among atoms changes fundamentally. This is very well established. Does that happen in single protein molecules too? If it does, then, even the existing faith of biologists that proteins vibrate only mechanically, that pure mechanical motion would dramatically change the electronic potential distribution in the protein molecule. Thus, the electronic property would change fundamentally. The final concern would be, even if it is established that protein undergoes electronic and mechanical resonance simultaneously, how does that matter biologically? The answer is that, apart from the thermal source inside a living body, there are electronic sources like membrane potential, which can also regulate the resonance vibrations along with the heat. It would mean that the explanations we used to provide to a biological process would have another electromagnetic world of communication squarely parallel to the chemical-only biology existing today.

Here we take four different biological proteins and image the electronic potential live for the proteins during thermal fluctuations and under artificial dc electromagnetic signal pumping. This shows that heat and electrical triggers generate similar spontaneous changes in the electronic potential distribution in protein and protein complexes. We applied high-frequency ac signals to proteins and protein complexes to find that at very specific frequencies, the proteins start conducting, which means the proteins turn transparent. At these frequencies, we have captured the electronic potential of the systems to find that a very particular kind of mechanical oscillations is generated. When we applied mechanical forces on the protein molecules using atomic sharp needles, we found that the electrical potential distribution all over the molecules and the protein changes. These findings suggest that proteins mechanical and electromagnetic resonance have one-to-one correspondence and if any one of them is changed, the other changes simultaneously.

2.7 Tubulin Resonance: Phase-Modulated Electric and Magnetic Field Profile Across Tubulin Dimers-Clocking Behavior

Using a tubulin dimer's actual structural data file, we have reported the phase-modulated electric and magnetic field profile across tubulin dimers [35]. We built tubulin structure theoretically and ran a dielectric electromagnetic simulation for obtained phase-modulated energy profile. We simulated the resonance frequencies and distribution of electric and magnetic fields in CST, observed the phase response behavior across αβ tubulin dimers at particular resonance frequencies, and found field energy's inhomogeneous nature. The resonance frequency shows a characteristic topological feature of tubulin dimer in such a way that, at a particular resonance frequency, the magnetic field dominates at a certain phase angle. In contrast, the electric field is dominated at a certain phase angle which leads to node and antinodes concept. Both fields distribution is the function of phase angle and shows the nature of clocking behavior over one complete phase cycle. Electric and magnetic fields generate the standing wave pattern along with a rotating field. Rotation of field or clocking is the fundamental of resonance property such as clock reside inside the clock.

2.7.1 Coherent Oscillations in Proteins

Electromagnetic resonance is the fundamental of information processing in biological molecules like DNA, RNA, and proteins and their complexes. The electromagnetic resonance occurs due to charge transfer through the biological molecules [76]. In a previous study, it has been reported that electromagnetic resonance occurs in biological molecules in a wide frequency range (KHz (10^3) to THz (10^{12})). The resonance of tiny biological molecules is very useful to measure molecular biological activity through computer simulation [77, 78]. Microtubule and tubulin show the electromagnetic resonance in kHz to THz frequency range [79]. Resonance occurs in tubulin or protein by charge transfer through protein molecules. Resonance frequency depends on the velocity of charge molecules from which we can speculate their resonance frequency range. Tubulin comprises two subunits, α sub tubulin and β sub tubulin, which are similar but not identical in the dimeric form [80]. α sub tubulin and β sub tubulin polymerize and make a long filament known as microtubule. The microtubule is an essential component for intracellular transport, movement, cell division in neurons. MT and tubulin are both macromolecules of our brain cells which are processing the information.

MT quantum state of de-coherence (order of de-coherence should be 10^{-13} s is important for neurophysiological effect [81]. De-coherence time is found slower in nature [82]. Taxol assists in restricting the dimensions of discrete microtubule bundles. Such microtubule assembly is more effective for cancer cell treatment because the combination of MTs is more stable and prevents the cell division process, resulting in stopping cancer cell growth. MTs have functions like preserving cell geometry, isolation of chromosomes, transport, and mobility of cells. The agents that affect the microtubule's equilibrium state are effective anticancer drugs, are useful for clinical use, and can destroy tumor vasculature [83]. The most valuable group of these agents are taxanes, vinca alkaloids, and epothilones, which have promising results in clinical trials [84]. Tubulin molecules are self-assembled in a linear chain and form a 2D sheet. Self-assemble phenomena could be observed by the atomic sharp tip of the coaxial probe [35]. The geometry of biological components is key to understanding the resonance-induced information processing in the brain.

A single microtubule has a strong electric dipole moment. Vibrations are expected in microtubules due to tubulin heterodimers' oscillating behavior and the spiral symmetry of the protein arrangement. Tubulin heterodimers act as electric dipoles, the resultant electromagnetic field generated around the microtubule. The mechanical properties of αβ tubulin dimer are affected by the existence and nonexistence of applied electric field [85]. The calculated and measured electromagnetic fields depict the electric field profile across the microtubule when tubulin heterodimers are excited at 1 GHz. Parameters like elastic constant and young's modulus of microtubule define its functional properties [86]. Diamagnetic susceptibility of the tubulin dimer is considered by axial symmetry [87]. We have reported the electric and magnetic field across αβ tubulin dimer when both sub tubulins are excited in the THz frequency domain. We considered only two parameters for simulating the field outside the microtubule: resonance band and em energy distribution over tubulin. We have built the dielectric model of tubulin (includes α and β tubulin), simulated the model, and got the resonance band in the THz frequency range. Energy sources are applied at one end or both ends of tubulin and analyzed phase-modulated electromagnetic field around tubulin at various resonance frequencies. To find the significance of phase-modulated electric and magnetic fields over the tubulin, we performed a simulation in various steps varying field conditions.

2.7.2 Tubulin Simulation Result

Step 1—We simulated tubulin structure by considering the whole tubulin as single material. The resonance frequency has an oscillatory nature. Electric and magnetic fields have a noisy profile through the tubulin. The energy source gets resonance instead of tubulin.

Step 2—(i) To get the true structural resonance peaks of tubulin, we edited the tubulin PDB structure file and removed noise to make it suitable to run in the CST simulator that solves Maxwell's equations. We assigned the different components of tubulin by different dielectric materials. We got the sharp multiple resonance peaks by putting the energy source at one end of the tubulin subunit (α tubulin or β tubulin). When there is no electric and magnetic field distribution, only the waveguide port blinks which means the tubulin structure does not get resonance. (ii) We stimulated both subunits of tubulin such as α tubulin or β tubulin. Multiple peaks are found in the THz frequency domain. At all resonance peaks, electric and magnetic fields get maximum along the line of energy source. Energy does not transmit to the second tubulin subunit. The Tubulin model consists of a beta sheet and connecting wires in the spiral form that equally contribute to the multiple resonance peaks. Both fields are almost similar to case (i). The maximum magnitude of energy concentrates along the direction of the energy source.

Step 3—In this step, we cleaned the tubulin structure by reducing additional connecting wires and then simulated the resultant tubulin protein. The energy source is applied at one end of the tubulin subunit. Magnetic field domination is more with two peaks in the THz frequency range. EM energy is concentrated with the different domains of the beta sheet. However, energy does not fully transmit to another subunit.

Step 4—In this step, we tried to stimulate tubulin by putting a coaxial atom probe at a different region of tubulin. The energy source is applied at a single subunit. Electromagnetic energy transmits from one subunit to another subunit of tubulin with minimum magnitude. The resonance band has multiple peaks. Nano atom probe measures the internal signals inside the protein-membrane by placing it on its surface. The detection of such signal with high resolution can be reordered from the cell membrane using the patch-clamp method.

Step 5—Without using the coaxial probes as an integral part of the protein structure, we stimulated both α tubulin or β tubulin. Electromagnetic field distribution with high magnitude appears at each resonance peak. Energy transmits from one subunit to another subunit through beta sheet and other constitutes. EM field is distributed in a particular order across the tubulin structure during complete phase rotation. Similar kinds of features are found across the axon and microtubule.

2.7.3 Electric and Magnetic Field Clocking of AIS, Tubulin, and Microtubule

When we measured the reflection and transmission parameters for a biomaterial, we found that the resonance frequency changed or shifted. Those transfer positions include time-lapse, oscillation time, and periodicity. Over that period, the electric and magnetic fields show switching properties, or it turns on and off several times and exhibits clock behavior. **Tubulin clock**: The electric and magnetic fields parallel switch on and off for the entire phase cycle. **AIS Clock**: Electric and magnetic fields are complementary to each other. When the magnetic field intensity is maximum, at those regions, the electric field is silent. The field distributions are opposite for the duration of the next phase. **Microtubule Clock**: Both fields show a clocking but with split nature or magnetic field appears with high intensity but electric field is absent in one phase rotation.

2.7.4 Biological Clocking Behavior

The biological clock is an organism's natural timing device involving proteins that can interact with other cells. Many questions remain unanswered in circadian neurobiology, like how many oscillators exist in our body and how they generate daily oscillation. Hypothalamus consists of a large number of "clock" neurons that drive our intelligence rhythm. The biological clock could be found in our behavior by the neuronal activity cycle. The circadian oscillator is not only there. Oscillator molecule's timing is still subject to discussion because when they communicate to the brain region, they appear different from previous [88]. Phase clocking is the universal property of proteins and appears uniquely [20]. The resonance frequency is associated with the quantized features. The information consists of resonance peaks within each peak [25, 26, 89]. We have seen almost every brain component see the clocking behavior. Tubulin protein, microtubule, axon also show the nature of clocking behavior (see Fig. 2.7). Electric and magnetic fields are distributed in

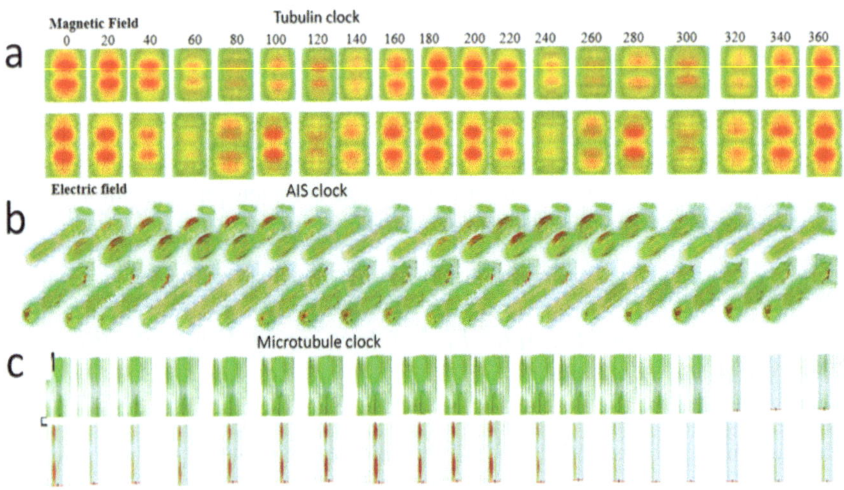

Fig. 2.7 Clocking of the electromagnetic wave along microtubule (**a**), axon initial segment, AIS (**b**), and microtubule (**c**)

various topological ways across these components. During one complete phase cycle, electric and magnetic fields get many times maximum and minimum intensity at one particular frequency. Both energies combine and create a standing wave pattern on the system (see Fig. 2.7). A standing wave is a part of the clock, and it is an integrated form of time and space mapping. At one particular mode, time and space, both are integrated. The protein clock is a combination of poly clocks.

The major finding of our study can be explained in terms of energy propagation across the αβ tubulin dimer. By exciting both tubulin dimers, energy is successfully passed. We have carried out the fundamental feature of tubulin dimer, such as clocking behavior. Such a feature is central in other proteins like actin, ankyrin, beta spectrin, clathrin, and snare complex are reported in the previous study. Such proteins show the properties of time crystals which are suitable for information processing. Such all proteins/crystals repeat their pattern periodically in time and as well as space. If we walk, any direction from any spatial point in crystal, we will find different kinds of atoms arrangements. Time crystals have hidden information inside them. We studied how a clock resided inside other clocks in tubulin dimer by considering standing wave patterns along with the rotating field.

2.8 Fusion of Evanescent Wave and the Tunneling Wave Function

An evanescent wave arising in the intricate geometry of a local structure diverges in all directions when it falls on a negative refractive index material, while converging the wave, it collects and groups those radiations into an organized coherent beam. The evanescent wave is mixed with the tunneling signal. Sharply converging a divergent beam may force the evanescent beam to focus on a point once inside the lens and outside [90]. The basics of quantum mechanics teach us that the signals produced from separate sources cannot generate entangled coherent waves. However, when an evanescent wave is born at the anisotropic material, then it entangles with the quantum wave functions tunneling through the site at that moment. They act as a single source . Thus, in a quantum cloaking experiment where quantum tunneling is measured in the presence of external electromagnetic wave exposure, the high-resolution features of the material hidden in the evanescent wave mix with the wave function. The tunneled matter wave contains the amplified version of the quantum dynamics. Thus, the quantum cloaking experiment is unique microscopy of quantum dynamics.

There are multiple ways of cloaking. One is the superlens concept introduced by Pendry. He said that if both electric permittivity and magnetic permeability are negative, the refractive index would be negative, which means the wave front would advance opposite to the direction of energy flow. The same phenomenon may cause the formation of superlens. This is the definition of metamaterial too. Pendry suggested that the evanescent waves that would normally decay away from the object in positive index materials instead grow exponentially through the NRIM, compensating for the decay in the rest of the optical path outside of NRIM. This allows the possibility of reconstructing a diffraction-free image by collecting all Fourier components, including these evanescent waves at the image plane. A superlens focuses all Fourier components from the object onto a two-dimensional image with a resolution far below the diffraction limit. The first evidence of evanescent wave in 35 nm silver film superlens was shown by Liu et al. [50].

The question is, how could resonant oscillation trigger faster and efficient tunneling? Garcia de Abajo et al. [91] made a 2D periodic array of 150 nm tiny holes on a 2D surface to create evanescent waves. The holes are filled using electromagnetic resonating spheres. Now the plotted classical transmittance across the device along with quantum tunneling as a function of frequency. The one-to-one correspondence showed that tunneling between electromagnetic resonance modes passes matter waves a long distance. Evanescent mode couples the resonance modes and holds them at a fixed value, not allowing them to decay exponentially. If the material does not have a negative dielectric constant, evanescent mode cannot take over and regulate the resonance feature. Therefore, the material would become invisible at certain frequencies, and at certain frequencies, they would turn visible.

2.9 Artificially Created Metamaterials and Nature Metamaterials-Based Electromagnetic Cloaking

2.9.1 Artificially Created Metamaterials-Based Electromagnetic Cloaking

This section will discuss how electromagnetic cloaking is carried out by developing the artificially created metamaterial device. Electromagnetic cloaking is an interesting research topic for researchers who are developing metamaterials in the form of artificially structured composite materials. When an electromagnetic wave passes through the electromagnetic metamaterials, the EM wave interacts with its structural features, and the diameter of the cloaked object is smaller than the EM wavelength. Microwave frequency metamaterial consists of an array of electrically conductive elements like loops of wire which have the inductive and capacitive property [92]. Photonic metamaterials are fabricated at the nanometer scale, and it manipulates em wave characteristics in the THz frequency domain. Electromagnetic cloaking is a device that makes an object invisible for a certain frequency range when electromagnetic energy is pumped into the object. An object appears invisible when em energy does not reflect from an object and does not scatter in any direction. In another sense, an object does not absorb energy and does not disturb the existing field outside the object [93]. The object is still present at the defined location, but the incident wave directs the object so that we cannot see it. So electromagnetic cloaking is the device that reduces the scattering cross-section of an object. So an ideal cloak removes all light scattering from the object. However, it only seems to be at a single frequency. About cloaking devices, several other concepts have been proposed in the literature. There are several methods like metal plate cloak, transmission line cloak, and coordination transformer cloak to achieve cloaking. Metal plate cloaks and transmission line cloaks work better than coordination transmission cloaks [94]. Pulse propagation inside a waveguide with a clocked metal object is used to demonstrate cloaking [95].

To understand the self-adaptive invisibility cloak for dynamic surroundings, we must focus on combining two disciplines, deep learning and metamaterials [96]. So the literature is fulfilled for realizing invisible cloaks using artificially created metamaterials. A hybrid invisibility cloak could be realized by integrating the wave tunneling functionality of zero-index materials and wavefront tailoring functionality of transparent metasurface. Metasurface-based invisibility cloaks are more reliable because of the less complex and thinner cloaking shells [97]. In smart metamaterial cloaking, a metamaterial device makes the object invincible by the electromagnetic wave and acquires its properties from the elastic deformation of the clock boundary. The elastic metamaterial is used to make a device with a broad operating frequency range (10–12 GHz) with lossless em energy [98]. So in literature, there are plenty of articles on artificially created metamaterial based on electromagnetic cloaking.

2.9.2 Natural Metamaterials-Based Electromagnetic Cloaking

In nature, finding a natural metamaterial is difficult. Thus far, no natural example existed. We need both electrical permittivity and magnetic permeability to be negative. Resonant ferromagnetic systems show negative magnetic permeability at resonance. Those inside a cloaking region would be blind and cannot see outside using the wavelength where it cloaks. This is beautiful, as it is a silence in complete isolation inside a cloaking material. Why not explore the quantum world inside? So, when a microtubule is cloaking at a particular frequency, the events happening inside cannot see what is outside. The noise propagating outside in a particular frequency domain cannot see inside. It is an isolated world inside.

Each microtubule with a particular length has a particular resonance frequency band where both the electric permittivity and the magnetic permeability are negative, i.e., the refractive index is negative. Then several resonance bands are there where the refractive index is positive. Note that theoretically, we measure those frequencies as dielectric resonance where the reflection and or transmission coefficients S11 and S21 drop largely in value. Therefore, two kinds of electromagnetic resonance are observed in a single brain extracted microtubule. One type of resonance frequencies where the reflection is more than the input, the microtubule could cloak in these frequencies. Here we get metamaterial-like responses exhibiting cloaking or superlensing. Another type of resonance frequency where transmission power is more than the input, the material acts as an antenna. Emission from the microtubule includes the evanescent waves; as discussed in the starting paragraph, there is no modulated classical wave but nested clocks or time crystals. By selectively choosing shining frequencies, one could isolate a particular part of the microtubule and at the same time read the written electron density of states in other parts of the microtubule as a time crystal. The combination of antenna and cloaking features enables writing, reading, erasing information as stored charges at suitable locations in the microtubule, and reading emitted electromagnetic vortices as time crystals.

2.10 Engineering of Water: Metamaterial Cloaking of Microtubule and DNA

We have isolated individual microtubule and DNA molecules, placed them on the HOPG substrate. When we take quantum tunneling images, at particular frequencies, the DNA and microtubule molecules disappear or vanish [4, 99, 100].

Water droplets arranged suitably could turn a material's surface invisible. Thus, far superlensing or cloaking ability of water has been shown only in the large surfaces; we show here that water molecules (Fig. 2.8a) bonded chirally turn a 2 nm protein invisible. Properly we cloak a nanowire called microtubule found in all Eukaryotic cells (Fig. 2.8b), as we turn it invisible in the dielectric resonance imaging at around 5–13 GHz. We adopted chiral vibrational sum-frequency generation (SFG) spectroscopy to find that microtubule and its constituent tubulin protein both templates the chiral spine of the superstructure of water. The hydrophilic amino acid Histidine, 2-Amino-3-(1H-imidazol-4-yl) propanoic acid, an amino acid residue, arranges helically around the tubulin protein, and since its imidazole part is a known water channel builder, thus forms a helical water channel as the SFG studies show that its NH bond and the OH bonds 3660 cm^{-1} vibrate together resonantly. The average distance between a pair of Histidine residue in a tubulin dimer is 0.5 nm ± 0.2 nm, fit to absorb the infrared signals. Thus, the helical spine of chiral water molecules absorbs the polarized infrared signals. For tubulin, the internal helical spine of beta sheet and

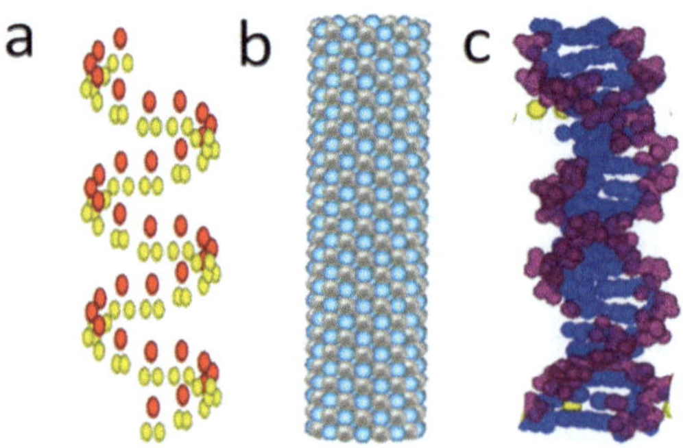

Fig. 2.8 **a** Helical shape of water layer; yellow ball-hydrogen, red ball-oxygen. **b** Microtubule structure. **c** DNA structure

external chiral water act in synchrony to convert it into a perfect lens. A microtubule has two water layers for its perfect lensing, as we see it live in the quantum tunneling images. Perfect lensing or cloaking protects the electromagnetic operation of microtubule and tubulin from ionic operations outside, i.e., lensing masks em biology from the ionic biology, known today, apart from generating a distinct time crystal.

In case of DNA (Fig. 2.8c), one could tune the optical properties by attaching the gold nanoparticles of different shapes and sizes in 2D and 3D form to DNA strands that could be extended or stretched as we need. The color of biomaterials could be changed, the advanced version of gold nanoparticles attaching to DNA strands could act as the cloaking device [101]. Nowadays, all metamaterials are made from some inorganic materials like copper and silicon, which show unusual properties of em waves. They allow them to respond as invisible cloaks, superlenses, etc. The metamaterial with unusual properties could be prepared using DNA as a basic unit. It happens by using polymerase enzyme to extend the DNA chains and put them non conventionally into the hydrogel. Resultantly, we get meta-hydrogel [102].

2.10.1 How Biomaterials Hold the Metamaterial Cloaking Property

The mass density of the first hydration layer of proteins and similar biomolecules is 15% higher than the bulk. Water molecules spend around a few picoseconds before leaving to the bulk. However, the first monolayer of water not just follows the molecule's pristine dynamics, it could make a stable spine of the molecule as a key integral component, e.g., DNA. We find that proteins like tubulin and microtubule act as metamaterial and exhibit cloaking in the presence of the water layer (Fig. 2.9). The inner and outer spine of chiral water shields the charge density distribution of a biomolecule from the diffusion of ions ready to neutralize. Since the coaxial probe records multiple time domains in a single event at the nanoscale, with an atomic resolution of 0.1 nm (a single atom tip) [103], we revisit an old hypothesis that a protein vibrates from milliseconds to femtoseconds periods like multiple clocks at a time [104]. To be functional, proteins are often considered to require a minimum hydration level of approximately 0.3 g of water per gram of protein, i.e., close to a monolayer of water [105].

Modulating the cylinder's core with non-magnetic conducting elements mimicking the water crystal located at the core of a microtubular structure. Water has a unique dielectric property [106]. We have already demonstrated that without the water channel, microtubule loses all its remarkable electronic features. Pentagon rings of water are fundamental to the microtubule core, but a similar arrangement of water is also fundamental to other proteins [107]. The water layer around the protein molecule is essential, and its dielectric property plays a vital role [108]. Protein hydration might affect wireless transmission,water molecules make an integral part

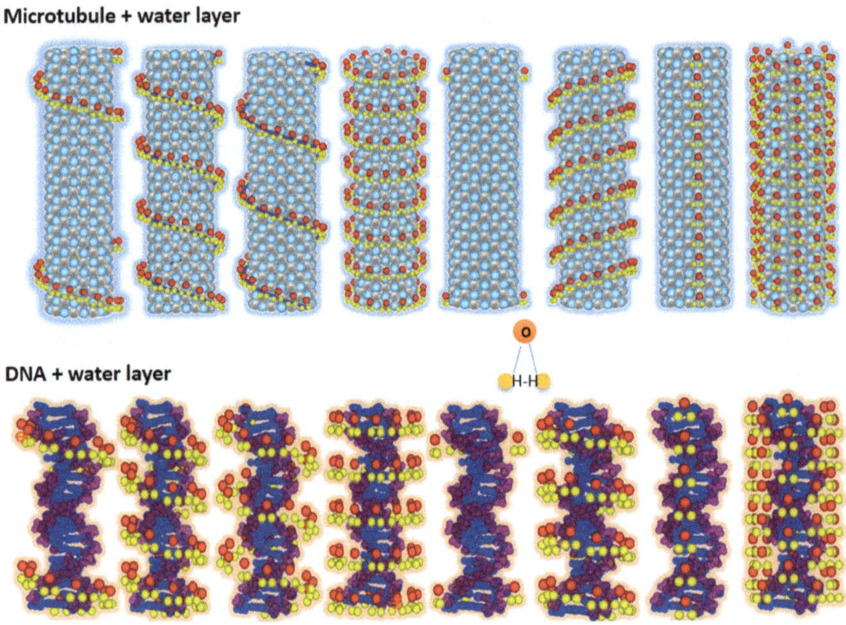

Fig. 2.9 The top panel shows the composition of microtubule–water, while the bottom row is about the DNA–water compositions. Different-shaped water layers are used to find the metamaterial properties of the microtubule and DNA

of the molecule [109]. Buried water molecules hold proteins' energy landscapes [110].

Since the 1930s, the dielectric resonance properties of a protein have been mixed with radio frequency induced dielectric heating of water. Ionic resonance corrupts the kHz band; no pure protein data exists. We introduce a new differential resonance measurement technique for isolating water's contribution from the protein resonance measurement data. Here we reduce water content on the protein surface <6%, hold on to that particular moisture, use a typically designed coaxial probe, wherein a differential sensing between a cavity resonance frequency and dielectric resonance frequency. We also add high-speed data transfer required for protein's real-time resonance measurements from 1 μHz to 178 GHz–0.2 THz. Two water layers, one inside the cylindrical hollow tube and the other surface above, ensure three nested clocks operating at a time in a single microtubule. A single microtubule processed a linguistic statement or argument like a computer chip in a single microtubule.

2.10.2 Resonance and EM Energy Distribution Along with the Composition of Microtubule-H_2O

In the 1960s came the culture of using a microlitre solution for resonance measurement. In the 2010s, we get the culture of nano-liter solution of proteins and various ways of confining fractions of proteins in the nano-cavity. Single protein measurements that started in 1988 using atomic probes are also improving with unique tips. One example is our coaxial atom probe [103]. Undoubtedly, minority researchers have succeeded in measuring the sharp resonance peaks of a protein. Those who succeeded used various novel tricks to avoid peak shifting methods that depend solely on the cavity's collective response. Despite a high Q value (10^5), coupled resonance measurement with background solution or cavity or both together is fatal. Calibrated data looks nice, but isolation goes through much intuitive manipulation in theory. One idea that we worked on was looking into a protein by reducing the water content, then using existing atom probes for three decades to measure the protein vibration directly.

To resolve this issue, we have put water molecules in the biomaterials and measured the effect of water in governing the resonance. Microtubule nanowire is found in every living cell of Eukaryotes. It is 25 nm in diameter and 25–30 μm in length. Inside is a hollow filled with crystalline water, and the upper surface is water too. We have estimated the resonance curve of this nanowire with and without water, using a model of microtubule in CST. Figure 2.10a is a water crystal just like the one we find inside a single microtubule. Then Fig. 2.10b is a microtubule alone with no water, then water inside the microtubule, both inside and outside (Fig. 2.10c, d). The resonance appears nearly similar (Fig. 2.10A), but electric and magnetic field distribution appear very different in all four cases (Fig. 2.10b). When we consider

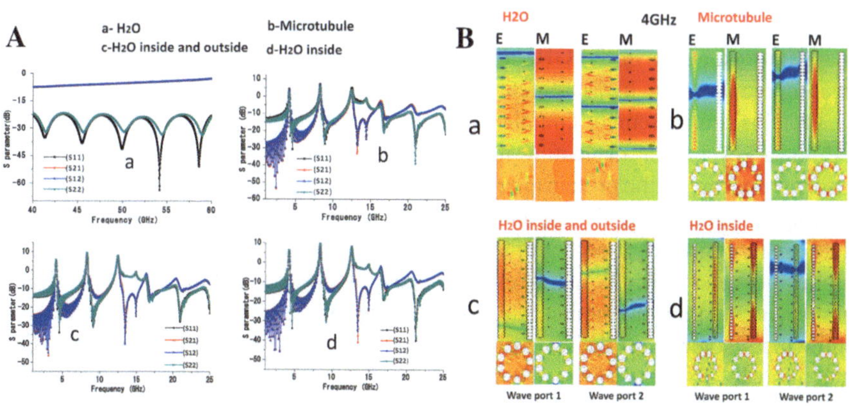

Fig. 2.10 **A** Resonance curves; reflection and transmission coefficients (S11, S12, S21, and S22) are produced by simulating the different compositions of microtubule and H_2O. **a**H_2O; **b** microtubule; **c** H_2O inside of microtubule; **d** H_2O inside and outside of microtubule. **B** Distribution of electric and magnetic fields are detected for the same microtubule and H_2O compositions at 4 GHz peaks

water both inside and outside, the electric and magnetic field starts clocking, and a spiral wave flows. Therefore, the water channel plays a key role in generating a clock, not the resonance peaks (Fig. 2.10B).

We theoretically study the exact role of water in and around microtubules and DNA. Energy is transmitted from one end to the other and reflected to the same port 1 along the chiral water in oscillatory form. In contrast, the energy source at the other end transfers energy at a constant rate in 40–60 GHz. The energy shifts to the low-frequency range (1–25 GHz), oscillating between ports 1 and2 and is reflected back (Fig. 2.10B) to the respective port. Chiral water superstructure wrapped around the microtubules in helical shape (Fig. 2.9). Compositions of microtubule–water structures oscillate in the same frequency range with high fluctuations. Energy is detected at 4 GHz along with the MW composition in vertical and horizontal view, as shown in Fig. 2.10B

2.10.3 Permittivity, Permeability, Refractive Index, and Impedance Parameters

Our body is full of water, 77%, so is protein and its complexes. Since the 1930s, high-frequency responses with and without protein solutions have been subtracted to measure protein's dielectric property. The basic protocol has not changed over 90 years. The entire food industry is based on detecting dielectric heating of water by radio frequency signal [111]. Since water transports ions, it is difficult to measure dielectric properties,because ions add noise in the low frequencies domains, biomaterials' true dielectric property is masked. On the other hand, in the GHz domain, the water molecules start resonating 5–90 GHz, water resonance blocks the reading, and then infrared emission of surrounding measuring elements takes over. For these reasons, most biomaterials are a few nm in dimension, supposed to show electromagnetic resonance in the GHz domain, but cannot show the real data. Dielectric heating regulates the measurement if the water content is >6–8% [112].

Proteins are extremely complex molecules. Although, the dielectric properties exhibit a key role in evaluating the functions of the protein. The dielectric resonance properties of the protein are mixed with the radio frequency, induced dielectric heating of water, the ionic resonance is corrupted in the KHz band, no pure protein data exists, or actual properties of the biomaterials are masked. In the GHz frequency range, the water molecule starts resonating and blocks the reading, and then infrared emits surrounding the biometrical. Most of the biomaterials in the nm dimension show the resonance in the GHz domain, but we cannot present the real clean data due to noise. To measure the real data, we have introduced the water layer with the biomaterial and theoretically studied the exact role of water in and around microtubules and DNA. Microtubule is the nanowire composed of the tubulin proteins with a core diameter of 25 nm and 25–30 μm in length found in every living cell. Here, we have estimated the resonance curves in terms of permittivity, permeability,

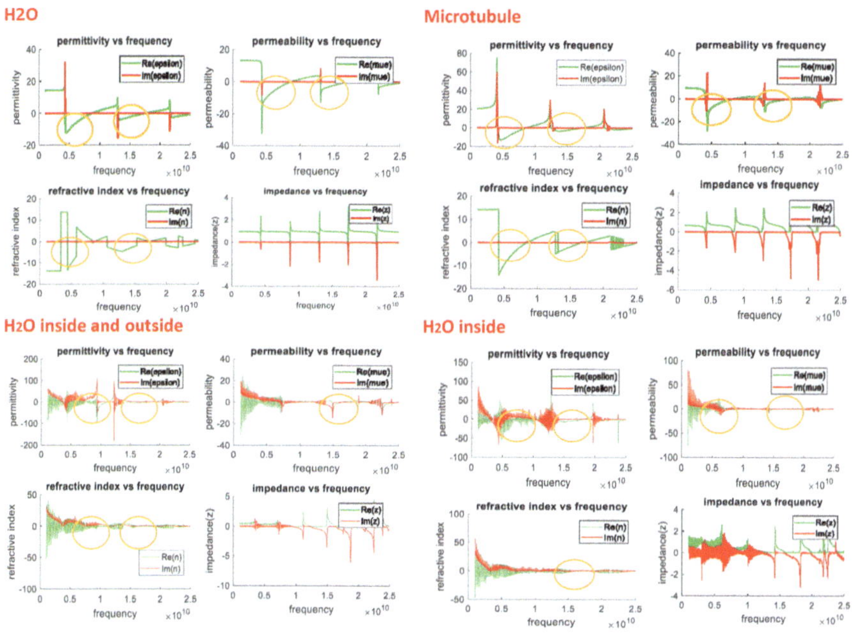

Fig. 2.11 Simulated permittivity, permeability, refractive index, and impedance curves are shown in the GHz frequency range for **a.** Single water crystal, **b.** Microtubule without water, **c.** Water inside and outside of the microtubule, and **d.** Water inside the microtubule

refractive index, and impedance for microtubules with and without water using a mimicked model in CST. It was proposed that water molecules within and out of microtubule cylinders have the characteristic of metamaterial with a negative value of the refractive index, impedance, permittivity, and permeability (see Fig. 2.11). When we consider the water with microtubule, it processes showing the property of metamaterial. Therefore, the water channel plays an important role in converting the biomaterial into metamaterial.

2.11 How Could Microtubule Show Cloaking?

Tubulin proteins have a negative refractive index, and the microtubule is a cylindrical array of tubulin proteins. Thus, it has a high probability of cloaking. Left-handed materials do not make a perfect lens due to absorption [113]. A periodic cascade of metamaterials could continuously amplify and eliminate the decay. A new concept was introduced called the plasmon injection scheme [114]. Metal's plasmon injection has an analog in the dielectric resonance of insulators. We have already been told that when the evanescent wave transmits at dielectric resonance, there is a coherent superposition. Imagine we are transmitting signals through tubulin proteins located on the microtubule surface. We activate the resonance modes of tubulin proteins. Since all the resonant modes are coupled along the transmission paths, the evanescent waves are produced across the entire lattice of the microtubule into a coherent beam that passes through the common energy band distributed spatially over the entire microtubule surface. Evanescent waves never acquire lossy modes. What we all need is the right externally applied auxiliary field. In the industry, engineers take metal like Cu, build resonators and assemble them in a helical or vortex pattern [51]. Nature has done the same thing for microtubules. It is a naturally found quantum cloaking device, similar to DNA, possibly the first one. Spatial periodicity is an important factor in eliminating the decay of evanescent waves [92, 115]. It means one could edit the microtubule surface lattice by applying a suitable resonance frequency, and then they would redefine periodicity, i.e., the transmission of evanescent waves. Tubulin protein array may act as a nanohole array [116].

Each microtubule with a particular length would have distinct frequency bands for cloaking and emission in a large cluster of microtubules of various lengths. Those microtubules whose cloaking frequencies are common would be invisible to each other remain silent when others communicate with that emission frequency. Thus, at a time, a large number of microtubules could communicate over a large distance. This is classical communication. However, the fact that even the water channel is visible at the top means all three layers could take part simultaneously or independently in the emission process as a single unit. Is this a classical communication or quantum?

The electromagnetic signal pumped into the system reflects, and we get much more signal than the input, and we saw these many times during the microtubule's electromagnetic resonance measurement. We reported that at particular ac resonance frequencies, a microtubule becomes transparent, its dc conductivity falls from 300 GOhm to a few MOhm, even hundreds of kilo-Ohm. Due to the formation of an evanescent wave, electrons could transmit through the central core.

There is an example of magnetic cloaking. At 21.3 MHz electromagnetic resonance, the magnetic field applied to a particular material passes through like a lens [51]. Each pixel of this lens was stacked layers of positive and negative refractive index materials. A consistent characteristic of the very near (evanescent) field is that the electric and magnetic fields are largely decoupled. This allows for nearly independent manipulation of the electric field with the permittivity and the magnetic field with the permeability [51]. In tubulin protein, microtubule, we observed isolation of electric and magnetic fields at resonance. It means that if we apply the electromagnetic field at resonance, the materials would spontaneously generate evanescent waves, which would change the tunneling probability. An array of metamaterial could hold the time-splitted waveforms of a much longer wavelength than its length. Microtubule thus holds or resonates MHz or 300 m of wavelength in a few micrometers length. Microtubule-like cylindrical superlens was proposed and studied extensively [117].

When we carried out tests for electromagnetic resonance studies for various biological materials (see Fig. 2.12; I–V characteristic measurement along microtubule with water). We found that electric and magnetic fields are isolated and move in a different route following distinct protocols. Decoupled electric and magnetic

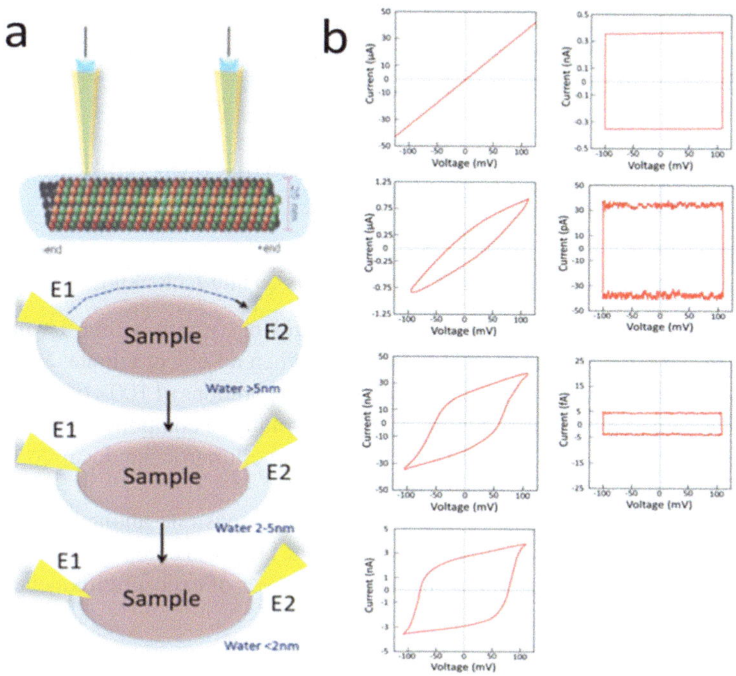

Fig. 2.12 **a** A schematic setup of I-V measurement along microtubule and water at different scales of water coverage using coaxial probes. **b** I–V characteristic along microtubule and water

fields depict the existence of an evanescent wave, which means an electromagnetic field that arrives there would create localized oscillations at resonance frequencies. Now, the question is why biological materials need evanescent wave manipulation? It turns out that evanescent wave enables lossless transmission, minimizing scattering [91].

Say we have a surface with subwavelength holes. As soon as the electromagnetic wave falls on the holes, the electric and magnetic vectors of the wave would split and distribute, adopting two distinct, peculiar symmetries. We get the generation of evanescent waves, i.e., localized electromagnetic oscillations. If we fill the holes with the resonators, then the transmission increases at resonance frequencies [91]. This is what biological structures do. They build dielectric resonators so in-homogeneously that their architecture anisotropy builds evanescent waves at those locations where biological systems make holes or resonating cavities. This is how to use a combination of cavity resonators and dielectric resonators. A biological system could design architectures to suitably localize the generation of evanescent waves at the desired location, whose local properties it would like to amplify and deliver to the others. The technologies for better imaging of biomaterials have been going on for a long time.

Using dielectric resonance microscopy, we imaged the microtubules inside the neurons [35]. Microfilaments do wonderful things in neurons [29, 118, 119].

2.12 Conclusion: Future Technologies Based on Anomalous Cloaking

There could be several applications of anomalous quantum cloaking.

1. Normally, it is believed that quantum tunneling occurs within a very small separation between two probes. However, when an object vibrates resonantly, then quantum tunneling could occur even at separations of 100 s of nanometers.
2. A large complex superstructure, if made of multiple components with distinct symmetries, then we could selectively observe a particular local component hidden deep inside a complex structure.
3. An assembly of objects with the ability to demonstrate anomalous quantum cloaking could build many coexisting circuits. Different components of different objects could be made visible by suitable choices of exposure frequencies.

References

1. Bandyopadhyay A (2020a) Nanobrain: the making of an artificial brain from a time crystal. Taylor & Francis Inc. Imprint CRC Press Inc., Bosa Roca, United States, p 336. ISBN 10-1439875499. ISBN 13-9781439875490. https://doi.org/10.1201/9780429107771.
2. Bandyopadhyay A, Ghosh S, Fujita D (2020b) Universal Geometric-musical language for big data processing in an assembly of clocking resonators, JP-2017-150171, 8/2/2017: World patent, WO 2019/026983; US Patent App. 16/635,900
3. Bandyopadhyay A, Ghosh S, Fujita D (2020c) Human brain like intelligent decision-making machine; JP-2017-150173; 8/2/2017; World patent WO 2019/026984; US Patent App. 16/635,892
4. Bandyopadhyay A, Fujita D (2021a) Method for realizing quantum cloaking in electromagnetic device for remote imaging apparatus. Application no 2021-172701
5. Ratner BD et al (2013) Biomaterials science: an evolving, multidisplinary endeavor. In: Biomaterials science, 3rd edn
6. Hieda, Nam KC (2013) Electric field measurement of the living human body for biomedical applications: phase measurement of the electric field intensity. Electromagn Biomed Appl 2013 |Article ID 305362 | https://doi.org/10.1155/2013/305362
7. Chen Y et al (2013) A feasibility study for microwave breast cancer detection using contrast-agent-loaded bacterial microbots. Electromagn Biomed Appl 2013 |Article ID 309703. https://doi.org/10.1155/2013/309703
8. Cattani C, Badea R, Chen SY, Crisan M (2013) Biomedical signal processing and modeling complexity of living systems. Computat Math Methods Med 1–2
9. Alon U, Surette MG, Barkai N, Leibler S (1999) Robustness in bacterial chemo taxis. Nature 397(6715):168–177
10. Barkai N, Leibler S (1997) Robustness in simple biochemical networks. Nature 387(6636):913–917

11. Kitano H (2007) Towards a theory of biological robustness. Mol Syst Biol 3:1–7
12. Kitano H, Oda K (2006) 2006: Robustness trade-offs and host-microbial symbiosis in the immune system. Mol Syst Biol 0022(msb4100039):E1–E10
13. Kitano H, Oda K, Kimura T, Matsuoka Y, Csete M, Doyle J, Muramatsu M (2014) Metabolic syndrome and robustness tradeoffs. Diabetes 53(3):S6–S15
14. Stelling J, Sauer U, Szallasi Z, Doyle FJ, Doyle J (2004) Robustness of cellular functions. Cell 118(6):675–685
15. Steuer R, Waldherr S, Sourjik V, Kollmann M (2011) Robust signal processing in living cells. PLoS Comput Biol 7(11):e1002218
16. Singh P, Doti R, Lugo JE, Faubert J, Rawat S, Ghosh S, Ray K, Bandyopadhyay A (2017a) DNA as an electromagnetic fractal cavity resonator: its universal sensing and fractal antenna behavior. In: Pant M, Ray K, Sharma T, Rawat S, Bandyopadhyay A (eds) Soft computing: theories and applications, vol 584. Springer, Singapore, pp 213–223
17. Singh P, Doti R, Lugo JE, Faubert J, Rawat S, Ghosh S, Ray K, Bandyopadhyay A (2017b) Biological infrared antenna and radar. In: Pant M, Ray K, Sharma T, Rawat S, Bandyopadhyay A (eds). Soft computing: theories and applications, vol 584. Springer, Singapore, pp 323–332
18. Singh P, Doti R, Lugo JE, Faubert J, Rawat S, Ghosh S, Ray K, Bandyopadhyay A (2019) Analysis of sun flower shaped monopole antenna. Wireless Pers Commun 104(3):s1-889
19. Singh P, Ocampo M, Lugo JE, Doti R, Faubert J, Rawat S, Ghosh S, Ray K, Bandyopadhyay A (2018a) Fractal and periodical biological antennas: hidden topologies in DNA, wasps and retina in the eye. In: Ray K, Pant M, Bandyopadhyay A (eds). Soft computing application, vol 761. Springer, Singapore, pp 113–130
20. Reddy S, Sonker S, Singh P, Saxena K, Singh S, Chhajed R, Tiwari S, Karthik KV, Ghosh S, Ray K et al (2018) A Brain-like computer made of time crystal: could a metric of prime alone replace a user and alleviate programming forever? Stud Comput Intell 761:1–44
21. Komal S, Pushpendra S, Pathik S, Satyajit S, Subrata G, Kanad R, Daisuke F, Anirban B (2020) Fractal, scale free electromagnetic resonance of a single brain extracted microtubule nanowire, a single tubulin protein and a single neuron. Fractal Fract 4:11
22. Pistolesi F, Manini N (2000) Geometric phases and multiple degeneracies in harmonic resonators. Phys Rev Lett 85:1585
23. Brinkmann K (1971) Metabolic control of temperature compensation in the circadian rhythm of Euglena gracilis. In: Menaker M (ed) Biochronometry. National Academy of Sciences, Washington, DC, USA, pp 567–593
24. Singh P, Saxena K, Singhania A, Sahoo P, Ghosh S, Chhajed R, Ray K, Fujita D, Bandyopadhyay A (2020) A self-operating time crystal model of the human brain: can we replace entire brain hardware with a 3D fractal architecture of clocks alone? Information 11:238. https://doi.org/10.3390/info11050238
25. Sahu S, Ghosh S, Ghosh B, Aswani K, Hirata K, Fujita D, Bandyopadhyay A (2013) Atomic water channel controlling remarkable properties of a single brain microtubule: correlating single protein to its supramolecular assembly. Biosens Bioelectron 47:141–148
26. Sahu S, Ghosh S, Hirata K, Fujita D, Bandyopadhyay A (2013b) Multi-level memory switching properties of a single brain microtubule. Appl Phys Lett 102(12):123701
27. Sahu S, Fujita D, Bandyopadhyay A (2015) US patent 9019685B2: Sahu S, Fujita D, Bandyopadhyay A (2010) Inductor made of arrayed capacitors (2010) Japanese patent has been issued on 20th August 2015 JP–511630 (world patent filed, this is the invention of fourth circuit element), US patent has been issued 9019685B2, 28th April 2015
28. Romanenko S et al (2017) The interaction between electromagnetic fields at megahertz, gigahertz and terahertz frequencies with cells, tissues and organisms: risks and potential. J R Soc Interface 14(137):20170585
29. Saxena K et al (2020) 2020 Fractal, scale free electromagnetic resonance of a single brain extracted microtubule nanowire, a single tubulin protein, and a single neuron. Fractal Fract 4:11. https://doi.org/10.3390/fractalfract4020011
30. Burr HS, Northrop FSC (1935) The electrodynamic theory of life. Q Rev Biol 10(3):322–333

31. Fröhlich H (1968) Long-range coherence and energy storage in biological systems. Int J Q Chem 2:641–649. https://doi.org/10.1002/qua.560020505
32. Blank and Goodman (1997) Do electromagnetic fields interact directly with DNA? Bioelectromagnetics 18(2):111–115
33. Jain SC, Tyagi K (1999) Effects of extremely low frequency electromagnetic fields on health. Indian J Biochem Biophys 36(5):348–351
34. Adair RK (1998) Extremely low frequency electromagnetic fields do not interact directly with DNA. Bioelectromagnetics 19(2):136–138
35. Sahu S, Ghosh S, Fujita D, Bandyopadhyay A (2014) Live visualizations of single isolated tubulin protein self-assembly via tunneling current: effect of electromagnetic pumping during spontaneous growth of microtubule. Sci Rep 4, Article number: 7303. https://doi.org/10.1038/srep07303
36. Slyusar VI (2009) Metamaterials on antenna solutions. In: International conference on antenna theory and techniques, Lviv, Ukraine, pp 19–24
37. Veselago VG (1968) The electrodynamics of substances with simultaneously negative values of and μ. Soviet Phys Uspekhi 10(4):509
38. Engheta N, Alù A, Salandrino A, Li J, Silveirinha MG, Edwards BE (2006a) From plasmonic nanocircuit elements to volumetric photonic negative-refraction metamaterials frontiers in optics. https://doi.org/10.1364/Fio.2006.Fmh2
39. Zouhdi S et al (2008) Metamaterials and plasmonics: fundamentals, modelling, applications, 2009. ISBN: 978-1-4020-9406-4. Part of the NATO Science for Peace and Security Series B: Physics and Biophysics book series (NAPSB)
40. Kshetrimayum RS (2005) A brief intro to metamaterials. IEEE Potentials 23(5):44–46. https://doi.org/10.1109/MP.2005.1368916
41. Engheta N, Ziolkowski RW (2006b) Metamaterials: physics and engineering explorations. Wiley-IEEE Press. ISBN: 978-0-471-76102-0
42. Eleftheriades GV, Iyer AK, Kremer PC (2002) Planar Negative refractive index media using periodically loaded transmission lines. IEEE Trans Microw Theory Tech 50(12):2702–2112
43. Soukoulis CM et al (2007) Physics. Negative refractive index at optical wavelengths. Science 315(5808):47–49. https://doi.org/10.1126/science.1136481
44. Fang N et al (2005) Sub-diffraction-limited optical imaging with a silver superlens. Science 308(5721):534–537. https://doi.org/10.1126/science.1108759
45. Pendry JB (2000) Negative refraction makes a perfect lens (PDF). Phys Rev Lett 85(18):3966–3969
46. Alù A, Engheta N (2005) Achieving transparency with plasmonic and metamaterial coatings. Phys Rev E 72:016623; Erratum Phys Rev E 73:019906 (2006)
47. Schurig D et al (2006) Metamaterial electromagnetic cloak at microwave frequencies. Science 314:977–980
48. Bullis K (2007) Superlenses and smaller computer chips. Technology review magazine of Massachusetts Institute of Technology
49. Fischer UC, Zingsheim HP (1981) Submicroscopic pattern replication with visible light. J Vac Sci Technol 19(4):881
50. Liu Z et al (2003) Rapid growth of evanescent wave by a silver superlens. Appl Phys Lett 83:5184–5187
51. Wiltshire MCK et al (2003) Metamaterial endoscope for magnetic field transfer: near field imaging with magnetic wires (PDF). Opt Express 11(7):709–715
52. Agrawal M, Vasyuchka VI, Serga AA, Karenowska AD, Melkov GA, Hillebrands B (2013) Direct measurement of magnon temperature: new insight into magnon-phonon coupling in magnetic insulators. Phys Rev Lett 111(1–5):107204
53. Henke W, Selzle HL, Lin SH, Schlag EW (1981) Effect of collision and magnetic field on quantum beat in biacetyl. Chem Phys Lett 77(3):448–451
54. Wang R, Zhang C, Zhang B, Liu Y, Wang X, Xiao M (2015) Magnetic dipolar interaction between correlated triplets created by singlet fission in tetracene crystals. Nat Commun 6(1):8602, 1–6

55. Welch DR, Genoni TC, Thoma C, Bruner N, Rose DV, Hsu SC (2012) Simulation of magnetic field generation in unmagnetized plasmas via beat-wave current drive. Phys Rev Lett 109:225002
56. Nye JF (2017) The life-cycle of Riemann-Silberstein electromagnetic vortices. J Opt 19(11):115002. https://doi.org/10.1088/2040-8986/aa8c41
57. Kim DH et al (2010) Biofunctionalized magnetic-vortex microdiscs for targeted cancer-cell destruction. Nat Mater 9(2):165–171
58. Im MY et al (2012) Symmetry breaking in the formation of magnetic vortex states in permalloy nanodisk. Nat Commun. https://doi.org/10.1038/ncomms1978
59. Richtmeyer RD (1939) Dielectric resonator. J Appl Phys 10(6):391–398
60. Choi JS, Howell JC (2015) Paraxial full-field cloaking. Opt Express 23(12):15857–15862
61. Yu N, Genevet P, Kats MA, Aieta F, Tetienne JP, Capasso F, Gaburro Z (2011) Light propagation with phase discontinuities: generalized law of reflection and refraction. Science 334(6054):333–337
62. Balanis CA (2005) Antenna theory: analysis and design. Wiley-Interscience, New York
63. Dallai R (2014) Overview on spermatogenesis and sperm structure of hexapoda. Arthropod Struct Dev 43(4):257–290
64. Gomes LF, Badke JP, Zama U, Dolder H, Lino-Neto J (2012) Morphology of the male reproductive system and spermatozoa in centris Fabricius, 1804 (Hymenoptera: Apidae, Centridini). Micron 43:695–704
65. Lanzavecchia S, Dallai R, Bellon PL, Afzeliuss BA (1991) The sperm tail of a gail midge and its microtubular arrangement studies by two strategies of image analysis(cecidomyiidae, dipteral, insecta). J Struct Bio 107:65–75
66. Zhang BB, Hua BZ (2017) Spermatogenesis and sperm structure of neopanorpa lui and neopanorpa lipingensis (mecoptera: panorpidae) with phylogenetic considerations. Arthropod Syst Phylogeny 75(3):373–386
67. Ishikawa H, Marshall WF (2011) Ciliogenesis: building the cell's antenna. Nat Rev Mol Cell Biol 12:222–234
68. Zheng X et al (2016) Molecular basis for CPAP-tubulin interaction in controlling centriolling centriolar and ciliary length. Nat Commun 16:11874 (1–13)
69. Kitagawa D, Vakonakis I, Olieric N, Hilbert M, Keller D, Olieric V, Bortfeld M, Erat MC, Flückiger I, Gönczy P, Steinmetz MO (2011) Structural basis of the 9-fold symmetry of centrioles. Cell 144(3):364–375
70. Witte H, Bradke F (2008) The role of the cytoskeleton during neuronal polarization. Curr Opin Neurobiol 18(5):479–487
71. Albrecht-Buehler G (1994) Cellular infrared detector appears to be contained in the centrosome. Cell Motil Cytoskeleton 27(3):262–271
72. Albrecht-Buehler G (1998) Altered drug resistance of microtubules in cells exposed to infrared light pulses: are microtubules the "nerves" of cells? Cell Motil Cytoskeleton 40(2):183–192
73. Aguirre EL (2012) Creating a 'Perfect' lens for super-resolution imaging. J Nanophoton 4(1):043514
74. Ghosh S et al (2016) Inventing a co-axial atomic resolution patch clamp to study a single resonating protein complex and ultra-low power communication deep inside a living neuron cell. J Int Neuro 15(4):403–433
75. Agrawal L, Chhajed R, Ghosh S, Ghosh B, Ray K, Sahu S et al (2016a) Fractal information theory (FIT) derived geometric musical language (GML) for brain inspired hypercomputing. In: Pant M, Ray K, Sharma T, Rawat S, Bandyopadhyay A (eds) Soft Computing: theories and applications. Advances in intelligent systems and computing. Springer, Singapore, pp 343–372
76. Matsukura F, Tokura Y, Ohno H (2015) Control of magnetism by electric fields Nat. Nanotech 10:209–220
77. Lottermoser T et al (2004) Magnetic phase control by an electric field effect. Nature 430:541–544

78. Stöhr J, Siegmann HC, Kashuna A, Gamble SJ (2009) Magnetization switching without charge or spin currents. Appl Phys Lett 94:072504
79. Cosic I et al (2015) Prediction of tubulin resonant frequencies using the resonant recognition model (RRM). IEEE Trans Nanobiosci 14(4):491–496. https://doi.org/10.1109/TNB.2014.2365851. Epub 2014 Nov 26
80. Tomasch WJ (1966) Geometrical resonance and boundary effects in tunneling from superconducting. Phys Rev Lett 16:16–19
81. Tegmark M (2000) The importance of quantum decoherence in brain processes. Phys Rev E 61:4194–4206
82. Hameroff S (2007) The brain is both neurocomputer and quantum computer. Cogn Sci (31):1035–1045
83. Hadfield JA et al (2003) Tubulin and microtubules as targets for anticancer drugs. Prog Cell Cycle Res 5:309–325
84. Pasquier E, Kavallaris M (2008) Microtubules: a dynamic target in cancer therapy. IUBMB Life 60(3):165–170
85. Havelka D, Cifra M (2009) Calculation of the electromagnetic field around a microtubule. Acta Polytech 49(2–3)
86. Saeidi HR, Lohrasebi A, Mahnam K (2014) External electric field effects on the mechanical properties of the αβ-tubulin dimer of microtubules: a molecular dynamics study. J Mol Model 20:2395
87. Bras W, Torbet J, Diakun GP, Rikken GL, Diaz JF, The diamagnetic susceptibility of the tubulin dimer. J Biophys 2014, Article ID 985082, 5 pages
88. Hastings MH (1997) Central clocking. Trends Neurosci 20(10):459–464
89. Agrawal l et al (2016b) Inventing atomic resolution scanning dielectric microscopy to see a single protein complex operation live at resonance in a neuron without touching or adulterating the cell. J Integr Neurosci 15 (04):435–462
90. Ziolkowski RW, Heyman E (2001) Wave propagation in media having negative permittivity and permeability (PDF). Phys Rev E 64(5):056625
91. Garcia de Abajo FJ et al (2005) Tunneling mechanism of light transmission through metallic films. PRL 95:067403
92. Shelby RA, Smith DR, Schultz S (2001) Experimental verification of a negative index of refraction. Science 292(5514):77–79
93. Alitalo P, Tretyakov S (2009) Electromagnetic cloaking with metamaterials. Mater Today 12(3):22–29. https://doi.org/10.1016/s1369-7021(09)70072-0
94. Alitalo P, Kettunen H, Tretyakov S (2010) Cloaking a metal object from an electromagnetic pulse: a comparison between various cloaking techniques. J Appl Phys 107:034905. https://doi.org/10.1063/1.3305322
95. Jing L, Zheng B, Xu S, Shen L, Chen H (2016) Experimental study on invisibility cloaks. In: 2016 IEEE international workshop on electromagnetics: applications and student innovation competition (iWEM). https://doi.org/10.1109/iwem.2016.7504885
96. Qian C, Chen H (2021) A perspective on the next generation of invisibility cloaks—Intelligent cloaks. Appl Phys Lett 118:180501. https://doi.org/10.1063/5.0049748
97. Chu H et al (2018) A hybrid invisibility cloak based on integration of transparent metasurfaces and zero-index materials. Sci Appl. https://doi.org/10.1038/s41377-018-0052-7
98. Shin et al (2012) Broadband electromagnetic cloaking with smart metamaterials. Nat Commun 3, Article number: 1213
99. Bandyopadhyay A, Fujita D (2021b) Electromagnetic device, magnetic and electrical vortex synthesis device and magnetic and optical vortex synthesis device. Application no. 2021-172702
100. Bandyopadhyay A, Sahoo P, Fujita D (2021c) Self-learning by information processing device and self-learning for information processing method. Application no. 2021-172703
101. Lin Q-Y et al (2018) Building super lattices from individual nanoparticles via template-confined DNA-mediated assembly. Science 359(6376):669–672

102. Lee JB et al (2012) A mechanical metamaterial made from a DNA hydrogel. Nat Nanotechnol 7
103. Singh P et al (2020) Reducing the dimension of a patch-clamp to the smallest physical limit using a coaxial atom probe. Progress Electromagn Res B 89:29–44
104. Hamm P (2008) Ultrafast peptide and protein dynamics by vibrational spectroscopy. In: Braun M, Gilch P, Zinth W (eds) Ultrashort laser pulses in biology and medicine. Berlin, pp 77–94
105. Rupley JA, Careri G (1991) Protein hydration and function. Adv Protein Chem 41:37–172
106. Buchner R, Barthel J, Stauber J (1999) The dielectric relaxation of water between 0 and 35 °C. Chem Phys Lett 306:57–63
107. Teeter MM (1984) Water structure of a hydrophobic protein at atomic resolution: pentagon rings of water molecules in crystals of crambin. Proc Natl Acad Sci USA 81:6014–6018
108. Ebbinghaus S, Kim SJ, Heyden M, Yu X, Heugen U, Gruebele M, Leitner DM, Havenith M (2007) An extended dynamical hydration shell around proteins. In: Proceedings of the national academy of sciences of the United States of America 104(52):20749–20752
109. Otting G, Liepinsh E, Wuthrich K (1991) Protein hydration in aqueous solution. Science 41:974–980
110. Denisov V, Peters J, Hörlein HD, Halle B (1996) Using buried water molecules to explore the energy landscape of proteins. Nat Struct Biol 3:505–509
111. Venkatesh MS, Raghavan GS (2005) An overview of dielectric property measurement techniques. Can Biosyst Eng 47(7):15
112. Ryynanen S (1995) The electromagnetic properties of food materials: a review of basic principles. J Food Eng 27(4):409–429
113. Garcia N, Nieto-Vesperinas M (2002) Left-handed materials do not make a perfect lens. Phys Rev Lett 88(20):207403
114. Sadatgol M, Ozdemir SK, Yang L, Guney DO (2015) Plasmon injection to compensate and control losses in negative index metamaterials. Phys Rev Lett 115(3):035502. arXiv:1506.06282
115. Smith DR et al (2003) Limitations on subdiffraction imaging with a negative refractive index slab (PDF). Appl Phys Lett 82(10):1506–1508
116. Huang FM et al (2008) Nanohole array as a lens (PDF). Nano Lett 8(8):2469–2472
117. Pendry J (2003) Perfect cylindrical lenses (PDF). Opt Express 11(7):755
118. Singh P et al (2021) Electrophysiology using coaxial atom probe array: live imaging reveals hidden circuits of a hippocampal neural network. J Neurophysiol 125(6):2107–2116
119. Singh P, Sahoo P, Saxena K, Manna JS, Ray K, Ghosh S, Bandyopadhyay A (2021) Cytoskeletal filaments deep inside a neuron are not silent: they regulate the precise timing of nerve spikes using a pair of vortices. Symmetry 13(5):821. https://doi.org/10.3390/sym13050821

Chapter 3
Life is an Engineering Marvel of Water: It's Water that Manages Noise to Synthesis Life

3.1 Introduction

3.1.1 The Magic of Water

Water is essential for our life. 60% of the human body is water. We consume almost 3 L of water per day. Our body parts are composed of the maximum amount of water. Our skin has 64%, brain and heart contain 73%, kidneys and muscles have 79%, and bones have 31% of water [1]. Water has the excellent ability to dissolve many substances in itself, which allows the body cell to use minerals and chemicals in many biological processes. Carbohydrates and protein, which we use in food, are transported by water in the bloodstream. To release the waste material from our bodies, water is important.

Water provides the shape, size, stability, dynamics, and functions of biomolecules. It is responsible for the stabilization and packing of the protein structure. Also, it plays a role in forming hydrogen-bond networks and electrostatic interaction [2]. Water defines the folds of secondary and tertiary structures [3]. The folding mechanism for several double helix proteins, including the water layer, is investigated at atomic resolution using REMD (Replica Exchange Molecular Dynamic) simulations. The protein folding mechanism is mediated by solvation, and water is squeezed out from the hydrophobic core after protein structural formation [4]. Water-mediated protein interaction is a feature to recognize that biomolecular sampling is done over a wide range of temperatures [5].

The non-pairwise-additive model of water–protein interaction improves the accuracy and efficiency of the predicted model of protein structure [6]. 3D form of water around the protein is observed using neutron diffraction methods and X-ray [7]. A detail about diffraction theory and applications for the biomolecular could be found in Ref. [8]. The bound water along the human interleukin-1 beta is found with magnetic resonance spectroscopy [9]. So biological water at the protein interface is important for their equilibrium structure, function, biomolecular recognition, and protein–protein interaction. Water dynamics at protein surface (Subtilisin Carlsberg)

P. Singh et al., *Biological Antenna to the Humanoid Bot*, Studies in Rhythm Engineering,
https://doi.org/10.1007/978-981-16-9677-0_3

was probed directly within femtosecond resolution [10]. The changes that occur in the protein molecules during the denaturation process provide more details about the structure and function of the protein [11].

3.1.2 Interaction of Water–Protein

Water is key for protein activity. Proteins need to have their surface covered by a monolayer of water to be biologically active. Water fills the gap, cavities and adjusts unsatisfied hydrogen bonding on the protein's surface [12]. How the properties of water are affected by the behavior of proteins is yet to be partially understood [13]. Literature is full of experiments and theories which reveal the relation between the dynamics of proteins and water [14, 15]. Some experiments show the water dynamic at the protein's surface that could measure the qualitative changes. Such changes may be the reason for the decay of protein activity [16]. Dynamic properties of water molecules in the hydration layer of protein could be determined by Nuclear Magnetic Resonance Spectroscopy, NMR [17]. Mobility and order of the water molecules could be studied by NMR. Neutron scattering is also used to measure intracellular water dynamics in human red blood cells [18]. Buried water molecules, hydration layer of water, and bulk water, these three kinds of water environment are present in globular protein [19, 20]. The relaxation rate of water protons is useful to provide a theoretical explanation of the experimental findings. Spin–lattice relaxation rates of water protons are observed by considering all possible sources of dipole contributions created from water protons [21, 22].

Water is considered an integral part of biomolecules, and its properties change by interaction with biomolecules [23, 24]. So interfacial water play a key role in understanding the protein–protein interaction. Analysis of statistical thermodynamic and molecular dynamic simulation tools are used to observe the dynamic of interfacial water. The hydrogen bond rearrangement is important for protein–water interaction [25]. In MD, we need to maintain the gradient of temperature and pressure to study the transport properties of molecules in nonequilibrium molecular dynamics [26]. Water–protein interactions are also important in finding the structure and functional properties of proteins in foods [27].

3.2 Dielectric Properties: Proteins and Water, Proteins in Water

3.2.1 Neighbors in an Undefined Boundary is the Interface of the Universe

There are several techniques to determine the electric characteristic of water. We could measure the complex permittivity of water using photonic crystals in terahertz frequencies. Photonic crystal resonator, PCR has a high sensitivity for sensing the molecules in the optical region [28]. The High Q factor of PCR does not suffer from an ohmic loss in the THz band. We determine the complex permittivity of water using photonic crystals in terahertz frequencies. We can also use the ultra-compact biochemical sensor. Transmission spectrum shows the shifting on the wavelength that strongly defines refractive indices of materials used to build ultra-compact biochemical sensors [29]. So, we measure the complex permittivity of bioliquid in the human brain in terms of frequency [30]. Terahertz absorption spectroscopy is used to determine the dynamic vibration of biological polymers in water [31].

The study of hydration isotherms may help understand the interaction between water and biomolecules. The dielectric permittivity of hydrated protein is measured at the microwave frequency by the microwave perturbation technique. The water bound to proteins beyond the first hydration layer is responsible for dielectric loss at the microwave frequency. Absorption measurement provides information of water activity in the secondary hydration layer [32]. So dielectric measurement at the microwave frequencies could help understand the difference between normal and virus-infected cells [33, 34]. They have mm differences in microwave absorption. For example, DNA in normal and virus attacked cells absorb the water molecules in different ways [35].

Frohlich-Kirkwood dipole moment fluctuation model is used to calculate the dielectric properties of the protein. Electric permittivity may vary by varying the behavior of functional groups in the protein which share total dipole moment. Several factors like ligands or water molecules, change in the solvent, temperature, and pH value may affect the dynamic of dipole-moment determining functional groups [36]. The dielectric constants of secondary and tertiary structures inside the proteins are reported in the literature, however, their optimal value is still unknown. The dielectric constant of protein is not constant. It is a complex function, informing about the properties of sequence and structure of proteins. The gaussian-based approach may deliver the dielectric constant throughout the structure by using the 3D structure of constituent molecules [37]. Three important observations have been made regarding the dielectric properties of protein solutions. At low frequencies, increment in dielectric constant depends on the polarity difference between solvent and protein molecules. At higher frequencies, such increment increases due to the amount and hydration of protein molecules. Critical frequency depends on the size and shape of molecules and solvent viscosity [38]. The studies of the dielectric constant of solution offer some additional knowledge of the physical chemistry of its substance.

To understand the behavior of proteins in the buffer solution, it is necessary to know the response of biomolecules. The dielectric constant and polarization of molecules change when they are isolated from the natural ecological condition. The time-dependent analysis measures protein solution activity in terms of time delay, relaxation time, and dielectric constant [39]. The dielectric behavior and dynamical transition (displacement of atoms in hydrated protein between two regions) of the hydrated protein are correlated. Electrostatic fluctuations have described the coupling between the protein–water interface and atomic motions. It can be understood by charge modulation inside the protein [40]. Water and protein dynamics are observed in water–protein solution in different compositions of protein and water by varying the temperature at 300 K [41]. Beyond 200 K, such behavior is represented by the collective motions of many bounded and non-bounded groups of atoms. Below this temperature, simple harmonic oscillations are a responsible factor. Such transition, known as glass transition, shows some similarities between "change in the dynamical behavior of protein" and "change in viscosity and some other properties of the liquid" [42]. Glass transition temperature has the board range for hydrated lysozyme (180 $\pm$ 15 K) and hydrated globular proteins ($\sim$160–200 K) [16, 43]. We can measure the glass transition temperature by Raman and Brillouin scattering.

3.2.2 Quadrupole Coupling: Gravitational-Wave from Supernovae to Proteins

Nuclei with quantum spin number >½ exhibit the quadrupole moment. A quadrupole could be considered a composition of two dipoles that do not couple to the symmetric field. However, in the presence of an asymmetric field, a force, say electric field gradient, will be applied on the quadrupole. Quadrupole moment means rotation of quadrupole along the axis. In a magnetic quadrupole, there could be two identical magnets parallel to each other, and the north pole of one magnet is in front of the south pole of the other magnet and vice versa. Such a configuration does not have any dipole.

The power transfer distance is limited. Efficient power could transfer if coupling coefficient/ loss rate >1. Misalignment between transmitter and receiver may lead to a loss in wireless power transmission. Power could transfer wirelessly through the magnetic quadrupole coupling in the high refractive index of the dielectric resonator. In a magnetic quadrupole mode of the resonator, the resonator offers better efficiency by minimizing the ohmic and radiation loss. The quality factor of the resonator increases randomly [44]. Resonant cavity mode of hollow metallic structure also offers efficient transfer power wirelessly. Here, the coupling is done between a large chamber-cavity and wire loop-receiver [45]. The WPT efficiency can be improved by using magnetic metamaterials to enhance the evanescent wave coupling. The magnetic metamaterial increases the coupling between transmitter and receiver, reduces power loss [46].

3.3 Fat and Water: Do Not Hate Fat, Beloved Just After Water

3.3.1 Brain is Full of Fat; It Does Not Make the Brain Unhealthy

Our brain consists of 60% fat and has the unique composition of fatty acid, DHA docosahexaenoic acid. DHA is a major brain fatty acid. Fatty acids are organic acids, have at least one carboxyl group (–COOH) and long chains of carbon atoms, which are joined by a double bond (unsaturated acid) or single bond (saturated acid). Phospholipids and triglycerides derive fatty acids. Polyunsaturated and monounsaturated fats are good for our health. They build cell membranes and sheaths surrounding the nerves. It is important for inflammation, muscle movement, and blood clotting. Abnormal metabolism leads to disease and health-related problems. Caenorhabditis elegans are used in functional studies of fatty acid and derived lipids of fatty acid. Understanding regulation and deregulation of fat synthesis are important to manage the obesity plague [47].

To find out the biological and behavioral process of fatty acid is difficult because of its special characteristic. During biosynthesis, fatty acid involves more than one enzyme, making it difficult to utilize genetic approaches. The fatty acid molecules are difficult to visualize by fluorescent probe [48]. Lipid metabolism plays a key role in understanding brain functions, but it is still a mystery in normal/pathological NSC (neural stem cell). The fatty acid suppresses NSC activity in Alzheimer's disease. Lipids store in ependymal cells in the Alzheimer's disease brain [49]. Diary fat stored in lipid droplets expands under normal or pathophysiological conditions (DE). Such lipids droplet expansion, DE, is caused by increasing triglycerides (TAG). DE is obtained in Caenorhabditis elegans under genetic and dietary regulation. The genetic defects occur in peroxisomal beta-oxidation due to size expansion of lipids droplets [50]. The genetic behavior and complex nervous system of Caenorhabditis elegans make it suitable to find out the energy balance in the animal system [51].

Brain involves 2 polyunsaturated fatty saturated acids like docosahexaenoic and arachidonic acid. They release from the membrane and turn into active molecules. Primarily both acids are esterified into brain phospholipids. How our brain takes polyunsaturated fatty acids is still unclear. Polyunsaturated fatty acids, PUFAs, regulated pathways can alter our brain level. However, a study using animal models is promising to target PUFAs metabolism with drugs or diet [52]. PUFAs exhibit a wide range of applications in biological systems. They can regulate the biological properties of membranes and form powerful lipid signaling molecules. A small genetic model like Drosophila melanogaster and Caenorhabditis elegans are used to understand the PUFAs and PUFA-derived signaling in early stages of their development like meiosis, fertilization, etc. [53].

The function of fatty acid could be explained on behalf of the immune system. The latest research on fatty acid shows it behaves like short term, long term, and sensory memory. Sensory memory is the perception of enterocytes and hepatocytes cells of the small intestine. These cells consist of a cytoplasmic lipid droplet cycle and short-term memory located in those. Short-term memory uses parallel processing and provides resources to long-term memory [54].

3.3.2 *Fat and Human Diet*

The human brain consists of 60% fat. Fatty acid molecules play an important role in our brain performance. Fatty acids are not synthesized in our bodies. They are obtained from dietary food. The omega-3 fatty acid is important for the brain growth of humans in fetal and postnatal duration [55]. Docosahexaenoic acid is the structural fatty acid in the retina and central nervous system, and it is crucial for brain development [56]. Alpha-linolenic acid and cis-linoleic acid are essential fatty acids for humans but are sourced from the diet. Fatty acids and their derivatives have application in clinical implications [57]. Functions of the glial membrane and neuronal are changed by changing the composition of polyunsaturated fatty acids in diets. Those functions are linked to brain function and the retina. Such compositions may alter by taking the diet that has the unusual ratio of linoleic acid to alpha-linolenic acid [58]. Fatty acids, omega-6, and omega-3 affect gene expression, and their ratio is also an important factor for health. From literature studies, it has been detected that human beings are involved with a diet where omega-6/omega-3 is equal to 1. In western diets, this ratio is 16/1, leading to various diseases like cancer, inflammation, etc. [59]. Lack of fatty acids during infancy leads to delay in brain development and trigger aging. There may be disorders in brain functions. Essential fatty acids play a significant role in the synthesis and function of brain neurotransmitters and molecules in the immune system. The brain needs a continuous supply of such diets for a lifetime [60]. The central nervous system in vertebrates is affected by dietary lipids. Dietary lipids are necessary for accelerating myelinogenesis in the rat brain [61, 62].

Studies regarding dietary docosahexaenoic acid and DHA on neurological function focus on the retina or visual system. The development of the central nervous system could be measured by auditory brainstem conduction times (ABCT). ABCT is very sensitive for identifying the diet effect on neurodevelopment [63]. Researchers are interested in finding the possible effects of PUFA to reduce the sickness and death rate due to degenerative diseases of the brain. Many studies have shown how an unhealthy diet is related to obesity, but how diet correlates neurological changes in the brain is still unclear. A high-fat diet contributes to irregularities in the hypothalamus. The hypothalamus controls metabolism and homeostasis. A high-fat diet alters the physical structure of microglial cells. Microglia are activated by changing their mitochondria which help to supply the energy from the food. It has been seen, Uncoupling Protein 2 (UCP2) is responsible for changing the size of mitochondria. UCP2 regulates the energy utilization of mitochondria, hypothalamus, and glucose homeostasis [64]. Ghrelin hormone also affects the brain by regulating neuronal activity. Ghrelin makes robust changes in hypothalamic mitochondria in mice that depend on the UCP2 [65].

3.4 Classical Cloaking and Effort to Create Its Quantum Analog

In a classical cloaking, the lights from an image physically split, move all around the object to rejoin and rebuild the back-image at the front [66], the object turns invisible. Creating its quantum analog is difficult because the coordinates of a matter wave do not move invariantly with time. In the last decade, several creative ideas have been proposed. To note a few, instead of spatial coordinates (x, y, z), the quantized potentials were taken as coordinates (Vx, Vy, Vz) of a wave function. It turns localized ripples of matter waves into a wavelike flow of probability current density that transports energy [67]. That wave does not displace but disperse. The modified wave packet then splits and rejoins, similar to a classical cloaking. Another trick was to reduce the scattering cross-section of carriers or their effective mass to zero. More is the core–shell layer, better would be the push–pull of carriers, and the scattering tends to zero [68]. Thus far, no experimental evidence has existed for quantum cloaking. Guiding a wave function through complex paths like a classical wave is not a pure quantum scenario; to cloak, entire matter wave may tunnel through, ignoring paths. Going beyond normal invisibility, could we select one of the multiple paths, visualize one component in a composite, vanishing all others? That would be inverse or anomalous quantum cloaking.

Here, apart from vanishing and resurfacing in quantum images, DNA codes selectively appear at specific electromagnetic frequencies. Similarly, three distinct dynamic systems of microtubule nanowire, diffusive ions on its surface, helical protein branches, and water core selectively appear in tunneling images. Adjusting frequency, one could even tune the dynamics of the water core deep inside the microtubule. Theoretical fitting of selective cloaking in DNA and microtubule suggested that constructive and destructive interference of resonating em signals produced by different dynamic regions couple with quantum potential. Aharonov and Bohm combined the classical electromagnetic wave and quantum potentials; here we observe selection rules in that combination. Classical and quantum potentials collectively select part of a cloaking object that would be collision-less and parts that would perturb the tunneling wave function. Thus, anomalous quantum cloaking delivers two unprecedented quantum technologies. First, one could silence parts of a quantum object, select paths of tunneling. Second, by packing wave function with the classical evanescent wave, transmit a wave packet beyond quantum limits.

3.4.1 *Quantum Visibility/Invisibility of a Partially Resonant Composite, PRC*

In 1994, Milton showed that multi-layered concentric cylinders act as a partially resonant composite, PRC, i.e., if the sum of dielectric constants of constituents is zero. It means some layers have a negative dielectric constant. The dielectric constant of the layer would have a phase $(-1+i\delta)$ which would lead to the unphysical singularity at certain frequencies $(\delta \to 0)$ at the junctions. Therein, the electromagnetic field would wildly oscillate, locally; i.e., the evanescent wave would produce. If the evanescent wave amplifies the matter wave, it leads to near-perfect tunneling [69]. Across a PRC, those layers should be quantum mechanically invisible. On the other hand, if the charge density changes in a resonating cavity or spiral cylindrical layer, then the matter wave would scatter—no cloaking [70]. However, the dynamics of resonators in that cavity are transmitted to the top layer [71], which should be read as a localized density of states in a quantum tunneling image.

3.4.2 *Experimental Study of Quantum Cloaking and Anomalous Quantum Cloaking*

To experimentally verify quantum cloaking, we put calf-thymus DNA solution and freshly reconstituted brain neuron extracted microtubule nanowire on a freshly cleaved HOPG surface. The substrate is placed in an air Scanning Tunneling Microscope (STM), and an antenna is placed ~ 5 mm away from the STM tip. The antenna is connected to a microwave and a radio wave signal source (Fig. 3.1 A). We scanned the surface for each radiation frequency to find the samples' frequency disappear in

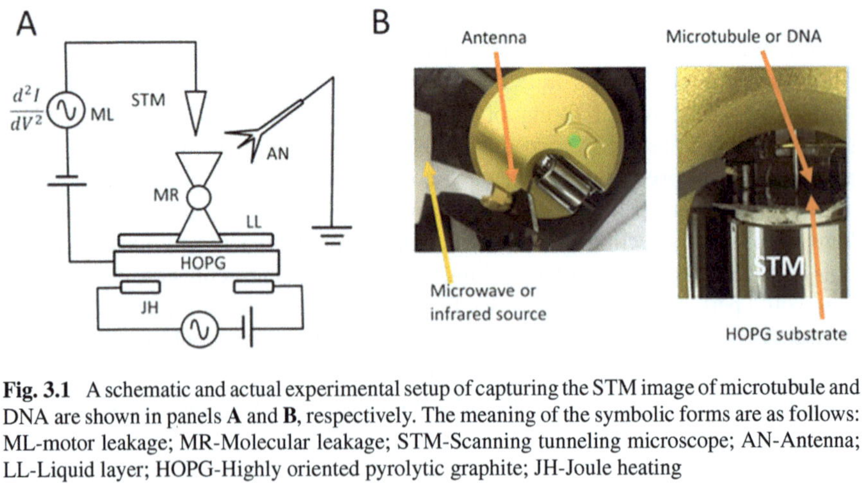

Fig. 3.1 A schematic and actual experimental setup of capturing the STM image of microtubule and DNA are shown in panels **A** and **B**, respectively. The meaning of the symbolic forms are as follows: ML-motor leakage; MR-Molecular leakage; STM-Scanning tunneling microscope; AN-Antenna; LL-Liquid layer; HOPG-Highly oriented pyrolytic graphite; JH-Joule heating

the tunneling image, and HOPG substrate at the bottom tunnels through the sample, reaching the top (tip) (Fig. 3.1B).

3.4.2.1 Three Helices of a DNA Molecule: DNA Turns Invisible

We carried out two studies on triple-helical DNA. Two helices are made of the molecule. The top helix is made of water. First, varied the antenna frequency at a gap of 0.001 GHz, i.e., high resolution, to find a magic frequency, but there is not. With increasing frequency, DNA gradually disappears, then, gradually, it turns visible. Second, to fit the selective appearance of DNA parts, we took an available database of variable number tandem repeats (VNTR) and generated an STM image theoretically. By matching experimental data of blinking DNA domains, we could read different classes of repeats, like a palindrome, mirror repeats, flanks in DNA codes. Therefore, by selecting suitable frequencies, one could reveal self-similar coding patterns in DNA. Most importantly, the revealed microsatellites and minisatellites of genetic codes suggest that silent domains in tunneling images are not inactive, rather act together with expressed codes.

Since we regularly capture microtubule images by applying MHz signals using an antenna [72], the tunneling image at GHz was a shocking observation, with no ions, no proteins. Instead, we observed a spiral, changing dynamics as a function of applied ac signal through the antenna. Microtubule has three distinct dielectric layers as three concentric cylinders, i.e., a bit more generic PRC than Milton class. Top ionic layer, middle protein layer, and at the center, there is a water channel. Water channel forms a microtubule core, but how 10–12 nm deep layer from any side is visible?

We set the antenna frequency to resonate at kHz so that the upper layer ions on the microtubule surface resonate. Then, the dynamics of adsorbed ions, primarily water molecules on the top layer, are captured in the quantum tunneling images. We set the antenna frequency at MHz to resonantly vibrate the tubulin proteins that make the central hollow cylinder. Then we observe the dynamics of tubulin lattice structures in the quantum tunneling images; new lattices are observed for each resonance frequency. Finally, we set the antenna frequencies in the GHz range by varying the resonance frequencies of the water channel; we visualize the dynamics of the localized density of states of the innermost water core. There were common resonance frequencies, and we could see the protein layer, water layer, and the surface at the bottom, which is clear evidence that matter wave penetrates directly through the material during quantum cloaking. Note that we should see the projection of image potential in a quantum cloaking, not the real water core or proteins.

3.4.3 Theoretical Study

We got back to theory. Modified microtubule structure in the simulator, studied two layers at a time, say only protein layer and water core, or protein and top ion layer alone, then finally studied all three layers together. By calculating the localization of electric and magnetic field distribution (Fig. 3.2), we found, only one layer could store the charge for a given frequency region; two layers remain silent. Milton studied only two-layered cylindrical structures [71]. There was only one junction with a negative dielectric material. Here, three layers in the microtubule could develop two negative dielectric layers at a time. Therefore, charge density distribution is near the lens where wave function would tunnel through, and evanescent wave amplification is modulated by device geometry. It means that, unlike DNA, for microtubule, two layers become a perfect lens at a time. Two evanescent waves amplify and make the signal loss nearly zero [70]. At the same time, two perfect lenses set two filters for matter waves to pass through. Fine-tuning of antenna frequency may vanish one layer, and we could see two other layers at a time.

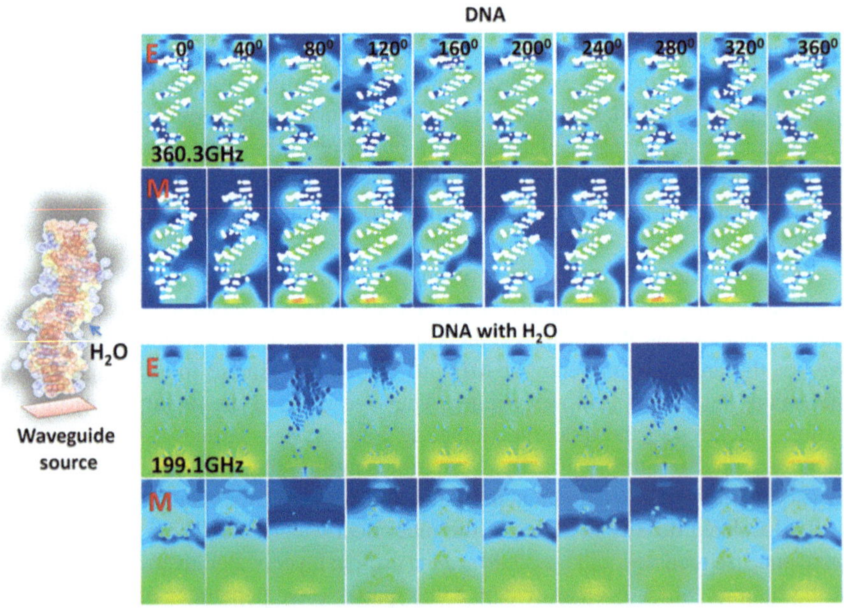

Fig. 3.2 Theoretical study of the electric and magnetic field of DNA and DNA + H_2O for one complete phase cycle (0–360°) is carried out at 30.3 GHz and 199.1 GHz resonance frequency, respectively. The waveguide port of suitable dimension is attached to one end of DNA and spirally wrapped H_2O layer (left panel). Electric and magnetic fields transmit along DNA without water layer (right top panel), but in the presence of the water layer, both fields are stored at the various functional region of DNA (right bottom panel) at the resonance frequency. E and M are represented as electric and magnetic fields

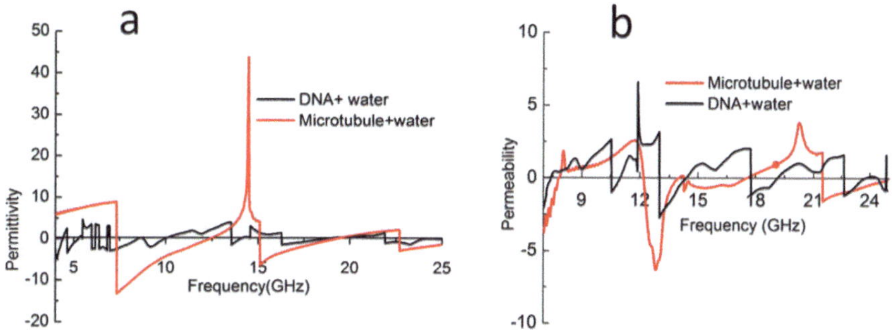

Fig. 3.3 Electric permittivity (ϵ) and magnetic permeability (μ) curves of DNA + H_2O and microtubule + H_2O are depicted in panel a and panel b, respectively. The compositions of DNA + H_2O and microtubule + H_2O are simulated in the EM simulator, providing the negative value of permittivity and permeability at particular frequencies for a given frequency region (0–25 GHz)

3.5 The Mutual Relationships Between Permeability and Permittivity

3.5.1 Functional Frequencies and Dielectric Fluctuations of DNA and Microtubule

In a PRC, the layer turns visible. If the material surface is anisotropic (e.g. a spiral), the dielectric constant turns negative at certain resonance frequencies. If both electric permittivity ϵ and magnetic permeability μ are negative in a medium ($\epsilon = -1; \mu = -1$), we get a superlens, or a perfect lens [73]. It means the re-positioning of a set of interconnected photons or pointing vectors on the other side of the lens. We theoretically created both DNA and microtubule structures as PRCs in a simulator (Computer Simulation Technology, CST) and solved Maxwell's equation to find the frequency range where DNA and microtubule act as a metamaterial (($\epsilon = -1; \mu = +1$); ($\epsilon = +1; \mu = -1$)), superlens, or exhibit anomalous dielectric resonance (Fig. 3.3). The matter wave may pass through when they become a perfect lens; if input frequency changes the charge density, the tunneling image will pick up that particular layer's activity.

3.5.2 Harmonic Nature of the Resonance Frequencies for Visibility/Invisibility of DNA and Microtubule

Remarkably, frequencies at which the water channel of DNA appears, microtubule's water channel disappears, and vice versa. Theoretical calculations show that the metamaterial properties of microtubule and DNA are complementary in two frequency domains. The percentage of the area visible for DNA and microtubule show that the complementary nature of the lensing effect or quantum cloaking emerges at least twice in the frequency domain. Moreover, we observe that the microtubule disappears at around 12 and 24 GHz. The next frequency would be 48 GHz. For DNA, vanishing occurs at 8, 16, and 32 GHz. It is like the harmonics of electromagnetic resonance frequencies (Fig. 3.4).

In order to understand the harmonics, we consider the geometric parameters, length, pitch, diameter, and lattice area of the spiral to build a phase space for allowed vibrations of a material. The spherical phase space has 12 holes or singularities. If one puts time function, say $\cos(t)$ in the spatial part (x, y, z), i.e., change geometric parameters with time, 12 holes open & close, one could count holes 0–12 (Fig. 3.5). The function coupled with the standard tunneling function gives

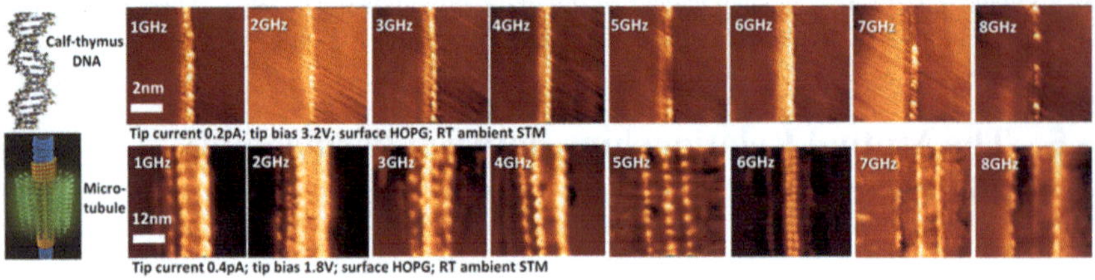

Fig. 3.4 Anomalous quantum cloaking of DNA (top panel) and microtubule (bottom panel) is measured in 1–8 GHz frequency region. The experimental setup is shown in Fig. 3.1. We have obtained an STM image on a period of 1 GHz frequency. DNA disappears at 8 GHz frequency. Microtubules begin to disappear after 7 GHz frequency, it completely disappears around 12 GHz

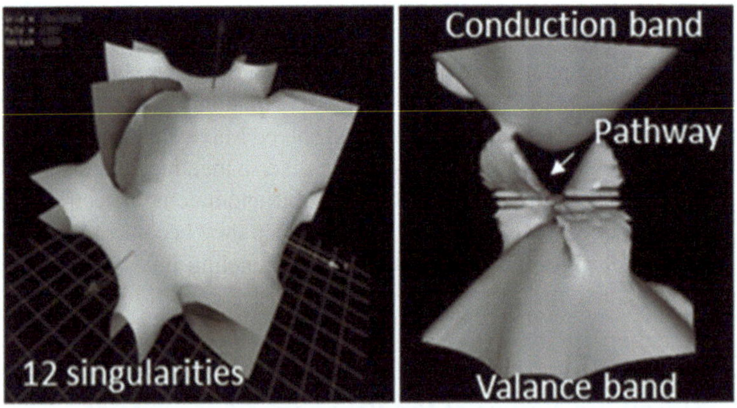

Fig. 3.5 12 singularity points in the Hamiltonian phase sphere are generated by varying length, pitch, diameter, and lattice parameters of a helical nanowire. We get such geometry by simulating $3(\cos x + \cos y + \cos z) + 4(\cos x \cdot \cos y \cdot \cos z)$ equation in a mathematical simulator, Mathmod 8.1. One bandgap of Hamiltonian is shown in the right panel. There are two bands; valence band and conduction band. One electron emits from the valence band and slightly touches the conduction band, and returns to the valence band. The electron generates a pathway in the bandgap; this happens instantaneously. In the next phase duration, the same thing happens but in a reverse way.

the tunneling images for all spirals studied here. The blinking of channels enables DNA and microtubule's water channels that are geometrically similar to behave in a complementary harmonics way.

3.5.3 Cloaking in DNA and Microtubule

Now, a material could act like a superlens at wide ranges of microwave and radio frequencies, not limited to the optical domain (e.g., 1.5 μm silver film). Superlensing in a core shell extends the tunneling path by ~5 times. It enables transferring a large packet of matter wave 30 nm apart—soon it would go much further. DNA is a two-layered, and microtubule is a three-layered Milton class PRC, clocking normally and anomalously, respectively. Aharonov–Bohm argued for directly adding the classical electromagnetic potential to the quantum potential of a matter wave. Here, in a multi-layered cylindrical material, the geometric relation between the dielectric materials decides how they would couple. Future studies in this direction might enable operating multi-dimensional quantum transmissions simultaneously. Anomalous quantum cloaking enables a live visualization of molecular dynamics of a part so deep inside a complex material at an ambient atmosphere, where, earlier, we could not even think of getting access from outside.

Quantum cloaking is not affected by the diffraction limit of electromagnetic signals, its angle of incidence, etc. Antenna pumped electromagnetic signals of various frequencies while scanning the tunneling current. Quantum cloaking in DNA and microtubule are shown in Fig. 3.4 top panel and bottom panel, respectivily, where matter wave created by constructive and destructive interference on the HOPG (matter wave on Graphene layer, Ψ_G) is retrieved across the material. We switched on and off DNA and microtubule repeatedly.

Since water is in the inner core, 12 nm above the HOPG atomic-flat surface where matter wave Ψ_G is located 12 nm inside the cylindrical protein surface of the microtubule, it is a challenge to retrieve its matter wave by quantum tunneling. 12 nm wide region below water layer and above it should act as a pair of lenses for both water's matter wave Ψ_{wat} and HOPG's matter wave Ψ_G. In both cases, DNA and microtubule, the frequency bandwidth for quantum cloaking is located a little higher frequency domain than classical cloaking Δf_m frequencies. Unlike classical cloaking, for quantum, an object appears/disappears for a wide frequency range.

We have theoretically regenerated the superposed matter wave for DNA and microtubule, suggesting that it is possible to bridge singularity in the phase space for energy transmission. Selective quantum cloaking requires a structure with at least one of the four periodic changes in length, pitch, diameter, and lattice area. We name this device Hinductor (H). DNA has two intertwined H devices while the microtubule has three. By applying a suitable ac signal, we open or close (blink) the singularity or hole (Fig. 3.5), thus selecting a composition of superlens in a composite of H that allows the tunneling of the matter wave. Each hole is associated with a structural symmetry. Since a hole has a boundary of allowed phase values, it represents a typical

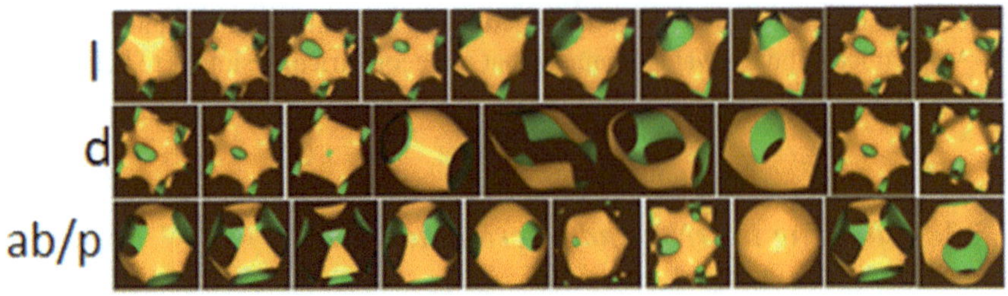

Fig. 3.6 The morphologies of phase sphere by varying the length, diameter, and lattice parameters/pitch of helical nanowire over a phase cycle are presented in top, mid and bottom panel, respectively. The equation of *H* is simulated in Mathmod 1.8, generates 12 singularities

value of length, pitch, diameter, and lattice area of H (Fig. 3.6). The closed holes of singularity domains allow local H structures with associated symmetries to print their typical matter wave on the generalized matter wave produced by the flat atomic surface.

However, the idea to build an elementary description of quantum cloaking begins from Schrödinger's equations [74]. The similarity between the Schrödingers equation and Maxwell's equation has been a subject of interest for decades. However, thus far, no reports exist on the experimental verification of quantum cloaking. One route could be taking quantum tunneling images of the object and showing them vanishing under the tunneling current images. Theoretically calculating the tunneling image is done by modulating the scattering function [75].

In a matter wave following Schrodinger's equation, theories of quantum cloaking suggested that the current of probability density behaves like an electromagnetic wave in Maxwell's equation. Two critical features of an electromagnetic wave, the displacement of electric–magnetic vectors and the field density (Poynting vector, *P*) hold their form even after a coordinate transformation. This invariance, a key to classical cloaking, required a quantum analog, so the current of probability density replaced the Poynting vector. A dispersion relation replaced the displacement vector. One problem with the existing Maxwell-Schrodinger analogy is that two parameters, the effective mass m^* and the potential V from isolated distinct parts of a matter wave, should bend and follow strict paths to bypass the material and rejoin similar to a classical cloaking. Moreover, entanglement in a matter wave demands ultralow temperature for a noise-free environment. For an electromagnetic wave, the anisotropic space and time ensure that distinct rays return to their original trajectory after bending through the object to hide, but its quantum analog is unclear.

We wish to find a general protocol to convert a material into a composition of superlens. Then, a matter wave could tunnel through the material at ambient atmosphere as it happens routinely in the quantum optics experiment through an optical lens. Not all materials are suitable for that. Milton argued for a unique material, where the elements in a composite collectively resonate, thus affecting far beyond their boundary. In spiral or vortex shape composites, we discovered a unique 3D phase space of energy transmission that enables selectively converting an element of the composite into a superlens and tunnel a matter wave from the flat atomic surface through that lens. Singularities or holes in phase space of transmission across a vortex are encoded in geometric parameters of helix or vortices that act as the quantum analog for the anisotropic space and time found in a classical cloaking. Tunneling through a superlens often extracts a 2D or 3D matter wave hidden deep inside a cloaking material, superposing it on the regenerated matter wave.

It is nearly impossible for an electron to tunnel through a 25 nm wide microtubule. However, we did it routinely for the last decade by applying an ac signal during tunneling measurement. We did not know the reason.

3.6 Theory for Anomalous Quantum Cloaking

3.6.1 *Quantum Tunneling and Scattering Tensor from Electron Density Field*

Based on the experimental data for the anomalous quantum cloaking of multiple helical nanowires, we have developed a theory for anomalous quantum cloaking; a simplified summary is presented here. We are to build an expression for the tunneling current image that maps matter wave profile by reading the local density of states. Thus, a quantum tunneling image is a replica of a matter wave packet on the surface.

A pixel on the matter wave is measured by a scanning tunneling microscope, STM the expression of tunneling current is $I(x, y) = \frac{e}{\pi\hbar}\int_0^{-eV} T_{st}(E+\mu(\upsilon))dE$, where the transmission function between substrate s, and tip t is$T_{\mathrm{st}}(E+\mu) = |S_{\mathrm{st}}(E+\mu)|^2\frac{V_i}{V_j}$, $S_{\mathrm{st}}(E)$ is the scattering tensor, $S_{\mathrm{st}}(E) = T_{\mathrm{sa}}S_{\mathrm{ab}}T_{\mathrm{bt}}$, relative group velocity between a pair of tunneling channels i and j is $\frac{V_i}{V_j}$. Here, a and b are two concentric cylindrical layers of a Milton class partial resonant composite, PRC. Here $T_{\mathrm{sa}}S_{\mathrm{ab}}T_{\mathrm{bt}}$ tensor product means tunneling from surface s to layer a (T_{sa}), then, scattering in the domain from a to b (S_{ab}) and finally tunneling from the measuring object a-b to the tip (T_{bt}). Scattering enables the matter wave of the surface to disappear and prints the atomic-scale dynamics of the measuring object. One has to neutralize the S_{ab} factor such that matter wave on the surface is printed in the scanned image by the STM tip.

3.6.2 *Hamiltonian for a Helical Nanowire*

In the presence of an electromagnetic wave, the composite layers undergo a resonant oscillation. STM image of a dielectric resonator changes significantly when an antenna pumps an ac signal with frequency υ wirelessly, we get $\mu(\upsilon) = \sum\Delta$. The AC signal amplifies tunneling current resolution, but it is not tip, but material resonates as a dielectric. Under ac exposure, not just the tunneling current, even the leakage dc signal transmission across a single microtubule nanowire is large. We failed to explain why microtubule conductivity turned 10^3 times more than its constituent tubulin protein. Resonance has measured the variations of transmission coefficient S21 and reflection coefficient S21 as a function of applied ac frequency. We observed a periodic change in the transmitted signal amplitude for different lengths, *L*; diameter, *D*; and the ratio of pitch, *P*, and lattice area ab. We repeated the experiment in a helical carbon nanotube, DNA, organically synthesized gel-based nanowires, and confirmed that even the dc transmission under electromagnetic exposure is a periodic function of *L*, *D*, and *P*/ab. Three blind experiments and theoretical fit for years confirmed that Hamiltonian for such a system is

$$H = \frac{\Omega^3}{2\pi^2}\left(\frac{2m^*}{\hbar^2}\right)^{\frac{3}{2}}\left(3\sum_i \cos\theta_i + 4\prod_i \cos\theta_i\right],$$

$\Omega = \frac{\text{Surface area}}{\text{lattice unit area}} = \frac{2\pi r L}{a \times b} = \frac{\pi \text{DL}}{\text{ab}}$, and m^* is the effective mass of the matter wave we want to pass through by cloaking. The function $3\sum_i \cos\theta_i + 4\prod_i \cos\theta_i = 3(\cos x + \cos y + \cos z) + 4(\cos x \cos y \cos z)$ is unique because if we replace *x*, *y*, *z* with Cost, periodic temporal oscillation shows that singularity or undefined regions open/close between 1 and 12, as shown in Fig. 3.5. Singularity means a set of *L*, *D*, and *P*/ab for which the vortex or helical system does not transmit. It bursts out energy from within or absorbs far from the frequency at which the system is pumped. Such a singularity burst shows negative S21, i.e., negative resonance. $S_{\text{ij}} = T_{\text{ik}} G_{\text{kl}} T_{\text{lj}}$, where $G_{\text{kl}} = \frac{\Omega^3}{4\pi^2}(3\sum_i \cos\theta_i + 4\prod_i \cos\theta_i)$, this is similar to a Green function, T_{ij} is analogous to the classical transmission coefficient S21.

3.6.3 *Determining Transmission Tensor Through Twelve Singularities*

The observed matter wave in the STM image $\Psi_{\text{mic}}(x, y) = \Psi_G + \hat{i}\Psi_{\text{ion}} + \hat{j}\Psi_{\text{wat}} + \hat{k}\Psi_{\text{tub}}$ is a superposition of four matter waves, a quaternion. Consequently, transmission through microtubule $\{T_{\text{mic}}(x, y)\} = T_{ij}$ is also a quaternion tensor. We extend Milton's formulation to estimate how one layer affects the neighboring composite layers. For example, one element of tensor would be $T_{ij} = \frac{V_{\text{ion}}(\mu_{s1})}{V_{\text{tub}}(\mu_{s1})}$, where $V_{\text{ion}}(\mu_{s1}) = \frac{\varepsilon_{\text{ion}} - \varepsilon_{\text{tub}}}{\varepsilon_{\text{ion}} + \varepsilon_{\text{tub}}}\left(\frac{r_{\text{wat}}^2}{r_{\text{tub}}}\right)^{2l} V_{\text{tub}}(\mu_{s1})$, and $V_{\text{ion}}(\mu_{s1}) = \frac{\varepsilon_{\text{ion}} - \varepsilon_{\text{tub}}}{\varepsilon_{\text{ion}} + \varepsilon_{\text{tub}}}\left(\frac{r_{\text{wat}}^2}{r_{\text{tub}}}\right)^{2l} V_{\text{tub}}(\mu_{s1})$. Similarly, 16 terms regulate one point in the transmitted matter wave. For a microtubule $\Psi_{\text{mic}}(x, y)$ is a quaternion. For DNA, it is a 3×3 matrix because $\Psi_{\text{DNA}}(x, y) = \Psi_G + i\Psi_{\text{nuc}} + j\Psi_{\text{wat}}$, DNA has a helical water channel.

Experimentally measured lossless transmission ($P_{\text{db}} > 1$) across 12 channels satisfy $\sum_{ij}^{12} |S_{ij}(E)|^2 = 1$, if only one channel is open, between two ports tip and substrate, and the rest 11 channels are closed, then, $|S_{11}(E)|^2 +$

$|S_{12}(E)|^2+|S_{22}(E)|^2 + |S_{21}(E)|^2 = 1$, Singularity channels only change phase by $\delta = \frac{n\pi}{4}$, n is the function of ac frequency. The scattering tensor is calculated by finding the reflection and transmission coefficients, $S11$ and $S21$, respectively. Using full-wave finite element simulation software (microwave studio, computer simulation technology, CST), we recreated tubulin protein, its complex microtubule, and DNA, simulated the S parameters ($S11$, and $S21$), plotted magnetic permeability, μ and electric susceptibility, ε. The plot of μ and ε as a function of frequency (Fig. 3.3) shows that at multiple ac frequency domains Δf_m, both microtubule and DNA turn into a metamaterial.

By solving Maxwell's equations for the microtubule constituents, together and separately for tubulin layer Tu, central water channel Wc, and ionic channel Ic on top; various compositions Tu + Wc, Wc + Ic, Tu + Ic, and Wc + Tu + Ic showed that both electric and magnetic fields are exchanged between elements synchronously around Δf_m. Dielectric constant ε is a function of electromagnetic frequency, we theoretically simulated and measured ε for isolated structures and found those frequencies at which microtubule satisfies $\varepsilon_{\text{ion}}+\varepsilon_{\text{tub}} = 0$, and $\varepsilon_{\text{wat}}+\varepsilon_{\text{tub}} = 0$. Reverse sign of dielectric constant when coupled together ensures core property (H_2O layer) extended to the outermost layer. The transmission path through the singularity channel is given by $S_{\text{st}}(E) = T_{\text{sa}}S_{\text{ab}}T_{\text{bt}} = T_{\text{sa}}T^{\text{ion}}_{\text{ar}}S^{\text{tub}}_{\text{rs}}T^{\text{wat}}_{\text{sb}}T_{\text{bt}}(\text{seetub}) = T_{\text{sa}}S^{\text{ion}}_{\text{ar}}T^{\text{tub}}_{\text{rs}}T^{\text{wat}}_{\text{sb}}T_{\text{bt}}(\text{seeion}) = T_{\text{sa}}T^{\text{ion}}_{\text{ar}}T^{\text{tub}}_{\text{rs}}S^{\text{wat}}_{\text{sb}}T_{\text{bt}}$. Here, we have expanded $S_{\text{st}}(E)$, for microtubule's three layers, we can expand further $S^{\text{tub}}_{\text{rs}} = T^{\text{tub}}_{\text{rm}}S^{\text{tub}}_{\text{mn}}S^{\text{tub}}_{\text{ns}}$ if we want to read, write-erase matter waves in a cluster of α helices in a tubulin protein (tub).

We design, synthesize, self-assemble helical nanowires for 12 layers one above another as tunable quantum cloaking devices to expand quaternion to octonion to a dodecanion tensor [76, 77].

3.6.4 3D Phase Invariant Structure of a Helical Symmetry

Group of stored charge Q diffused by noise combines the lattice spin waves (a and b are lattice parameters), by addition (like classical) and product (like in quantum) of phase $z = \left(\sum z_i + \prod z_i\right)$ as they beat locally ($z = e^{i\theta}$). Due to global topological constraints spread over three directions (diameter, D; pitch P; length L), local waves interfere to beat again.

We get for the beating of beats $e^{\delta^3 z} = e^{\delta^3\left(\sum z_i+\prod z_i\right)} = e^{\delta^3\left(3(z^\dagger_D+z^\dagger_P+z^\dagger_L)+4z^\dagger_D z^\dagger_P z^\dagger_L\right)}$.

Here, $\delta = \frac{\text{Surface area}}{\text{lattice unit area}} = \frac{2\pi rL}{a\times b} = \frac{\pi DL}{ab}$. The critical parameter to integrate classical and quantum factors in phase space should be in ratio 1:1.0205, i.e., 3:3.0615. Here, 3:3.0615 is the critical point where a singularity is born. We kept quantum capacitance induced phase factor 3:4 to keep the area of the hole created by singularity is proportional to the area covered by the phase space continuum.

The cumulative spatial phase factor of the charges $Qe^{\delta z}$ constituting the sinusoidal wave per unit velocity of photon ($Qe^{\delta^3 z}/c_m$) in that medium is the magnetic flux (since $B = E/c$, we get for entire cylindrical surface $\Psi = Qe^z/c_m = QH$).

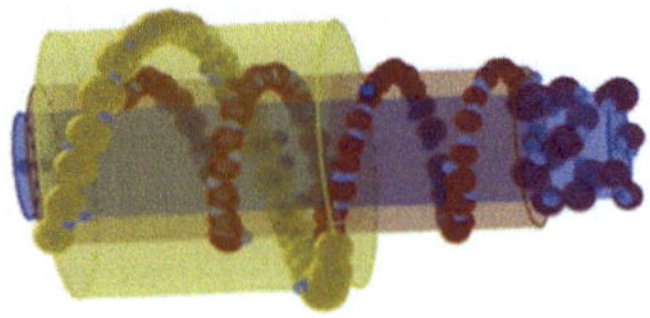

Fig. 3.7 An elementary *H* device with three helical nanowires is composed of balls of the insulator and metallic cylinder tubes

Here, $\frac{\Psi}{Q} = H = \frac{e^{\delta^3 z}}{c_m} = \sqrt{\mu_m \epsilon_m} e^{\delta^3 z} = \frac{1}{C_m} e^{\delta^3 \left(3(z_D^\dagger + z_P^\dagger + z_L^\dagger\right) + 4 z_D^\dagger z_P^\dagger z_L^\dagger)}$. The nth period plays an important factor in the magnetic wave. The larger the n, the phase oscillation gradient increases nearly exponentially.

$$H = \frac{1}{C_m} e^{\delta^3 \{3\cos(\frac{D-\mathrm{nb}}{b}) + 3\cos(\frac{P-\mathrm{na}}{a}) + 3\cos(\frac{L-\mathrm{nP}}{P}) + 4\cos(\frac{D-\mathrm{nb}}{b})\cos(\frac{P-\mathrm{na}}{a})\cos(\frac{L-\mathrm{nP}}{P})\}}$$

3.7 The Magical Twelve Dimensions and Twelve Singularities

Blinking of 12 singularities could happen in 2^{12} ways. If a hole or a group of holes periodically blink, it acts as a clock, and a clock is itself a memory state. In a complex matrix of *H*, the geometric orientation of *H* decides the partial contribution to Milton's collective resonant communication. We envision a supramolecular matrix of *H*, a jelly where zillions of *H* devices (Fig. 3.7) open or close their lenses to build the superposition of circuits in an ambient atmosphere. Only those modes which communicate are visible to each other, and the rest turn invisible. In the jelly of *H*, the phase spaces of zillions of *H* devices blink. In the holes, some other *H* devices can enter, self-assemble. Following Milton-Mansfield's argument, it is possible to replace an element H in a composite. Thus, a jelly could rewire, learn by plasticity. The concept of 1D–11D is depicted in Fig. 3.8. Space, time, topology and prime preceptions change by addition of dimension. For an example 1D–3D-spatial dimension; 4D–6D-time, 7D–9D-time; 9D–11D-prime.

3.8 Difference Between Classical and Quantum Cloaking

In classical cloaking, all electromagnetic waves bend together around a material and return to the original trajectory at a time; thus they rebuild the background image on its front side (Fig. 3.9, left). In quantum cloaking, a material has to break and recreate a packet of quantum wave function Ψ on the other side of a material at a time (Fig. 3.9, middle). Thus far, it was merely a theory. Here we demonstrate that using a suitable electromagnetic frequency, we can tune which part of the material would be

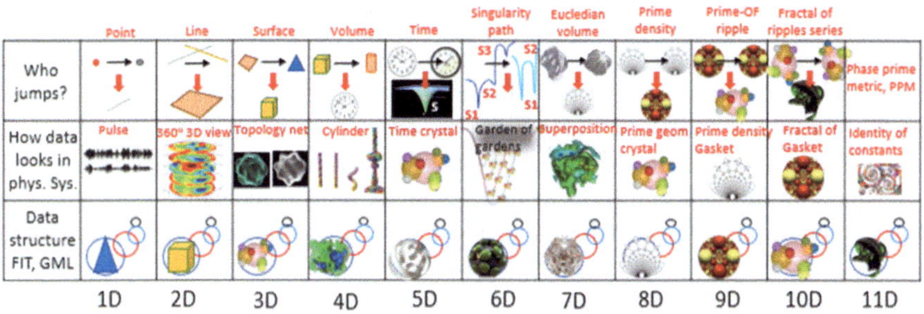

Fig. 3.8 In this figure, the first row deals with "asking questions" that generate the concept of 12 dimensions. The second row presents how the data may appear in a real physical system for those 12 dimensions. The third row is about how data structure looks like in FIT, GML [78–80].

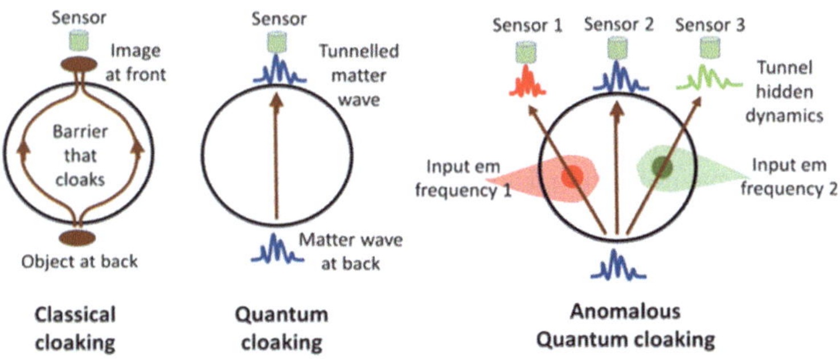

Fig. 3.9 A schematic of classical cloaking, quantum cloaking, and anomalous quantum cloaking is shown here. The object is represented by a circle. In classical cloaking, the incoming light splits from the object, travels around it, and is rejoined to form the image of object in front. The object becomes invisible. In quantum cloaking, the object could hide into superlens by pumping the EM signal of a suitable frequency. In anomalous quantum cloaking, object geometry can be tuned by applying the EM signal, and singularities/holes can be opened/closed in transmission. We can reproduce the matter wave on the other side of object

quantum mechanically invisible or visible (Fig. 3.9, right). We repeatedly vanished DNA in the tunneling regime and brought it back. Microtubule, a key nanowire of the cytoskeleton was vanished part by part using suitable frequencies. It has three distinct dielectric regions, the top ion-water layer, central tubulin protein layer, and the inner water core. We could image one part while vanishing the other two. Our quantum cloaking is anomalous as we can let the user see a tiny part's quantum resonance dynamics hidden inside a giant architecture.

A classical cloaking material at a time bends an entire packet of light falling all around an object's boundary. While bending, it keeps the relative coordinates of incident lights and transmits them at the same time as is, thus, making the object invisible. To return to the original trajectory, each ray of light bends differently. Hence, the bending space needs to be anisotropic. In a classical cloaking, the time-independent transformation of electric displacement vector and field density or Poynting vector ensures that the coordinates before bending and after rejoining remain intact. Quantum analogs of these two criteria, anisotropic space and time-independent transformation of coordinates, are required. In a quantum cloaking, the matter waves Ψ breaks and recreates on the other side of the object to hide. Here, wave function Ψ has to split into a group of matter waves, bend and join as one after passing through an object to hide its breaking and recreating entanglement. Since quantum transmission does not follow any defined path or canals of minimum energy, the demand to bend matter waves for the existing quantum cloaking theories similar to a classical electromagnetic wave is a trivial criterion.

When we consider matter wave tunnels as a whole through a material, it does not split or bend like a classical wave. A tunneling image contains a map of the local density of states. If an atomic-scale object is kept on an atomic-flat surface, one could vanish an object from the tunneling image by setting the right condition. The flat substrate surface would be visible alone. By pumping electromagnetic waves at the resonance frequency, we can create a phase singularity domain in the material, acting as lensing of electromagnetic waves. For us, the quantum tunneling image of material is like an optical image. Therefore, the material should disappear in this image during cloaking, and the background surface should be visible. The electromagnetic resonance of proteins has been known since the 1930s, and we have been reporting the quantum tunneling images of resonance of proteins since 2013. However, during quantum tunneling experiments, the disappearance of molecules in the tunneling regime is often ignored as artifacts. The atomic-flat surface is what is only visible in the images. Here our setup is a normal scanning tunneling microscope attached to a state-of-the-art Yagi antenna. Wirelessly we pump ac signals to the molecule as we capture the tunneling current image (Fig. 3.4).

A microtubule is a 25 nm wide nanowire whose surface is covered with ions, water molecules, 8 nm wide tubulin protein dimers make a hollow cylinder inside a 17 nm wide hole, where ordered water channels exist. No one could see it, access it by spectroscopic means. We were imaging the microtubule while pumping it wirelessly at around 6 GHz. We failed to see proteins, or ion-water complexes, which were routine studies for us. By tuning frequency when we observed different density of states of water's helical channel, it was obvious that the composition of the wave of water molecules perturbed by electromagnetic resonance created different architectures. Those architectures were passing through proteins and making them visible in the quantum tunneling images. In the normal event of quantum cloaking, we should notice invisibility at certain frequencies, but here, it is visibility, so we term it anomalous quantum cloaking (see Fig. 3.9, right).

3.8.1 *Origin of Quantum Cloaking: Helical Arrangement of H_2O*

When the water channel's density of state's dynamics is visible, we repeated tunneling at MHz to notice proteins and the dynamics of its local density of states and ion–water dynamics of the cylindrical surface at kHz. By varying the ac frequency that the antenna triggers, it is possible to see live the critically hidden parts of a giant complex. The observation raised a few queries. First is a microtubule complex of at least three different metamaterials. We created a replica of tubulin protein and DNA in the software, built microtubules and DNA with and without water and ions (mimicked DNA geometry- Fig. 3.10), and confirmed that water enhances the metamaterial property. We did the Second, what is the origin of quantum cloaking, helical arrangement of water, or helical topology itself? Using water droplets, one could convert any material into a metamaterial. However, we found that inner water crystal and external water–ion complex enhance invisibility index, but alone a helix could generate invisibility.

A helical path quantizes the electron density of states. Thus, resonance acts as an electromagnetic lens, which converges the lines of forces of the electromagnetic wave along its spiral path. Pancharatnam argued that in the domain of space-time, the geometric phase accounts for a spiral path. Following these two observations, we have formulated a theory that if electromagnetic lensing happens along a helical path and those path parameters resemble the geometric phase path of clocking electronic density of states, then that spiral alone regulates the tunneling. All other spirals would disappear in a matrix of distinct spirals. The finding comprehensively generates the part by part vanishing features of the microtubule.

One key part of this spiral cloaking is that we might generate a reverse cloaking. Suppose we have a pair of helices out of phase, then one geometric phase would cancel the other, thus restricting the tunneling totally. It means we could vanish a

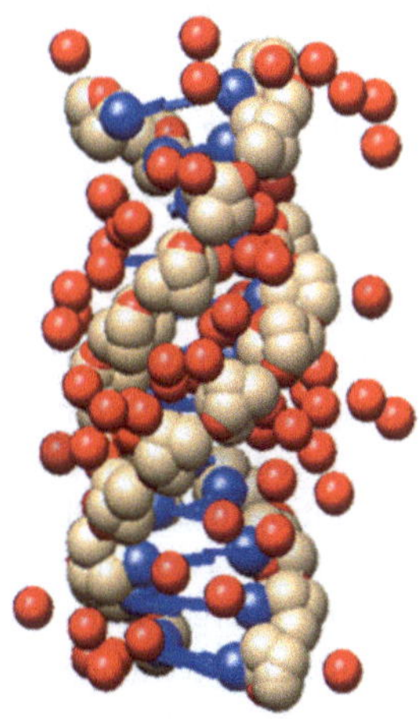

Fig. 3.10 A helical layer of H_2O around the DNA. We took the structural data file from Protein Data Bank and simulated it in CST. The red, white and green balls show water molecules, double helix and base pairs' molecules, respectively

double helix. To verify the prediction of our theory, we did the same experiment with DNA and could vanish it repeatedly from the HOPG surface (see Fig. 3.4).

3.8.2 *How Our Quantum Cloaking Phenomenon Differs from Earlier Prediction?*

In order to confirm that vanishing is real, we changed the pumping frequency using our antenna and found that quantum cloaking is not sudden magic. DNA disappears and appears as a function of frequency gradually. Most importantly, the invisibility of microtubule water channel and DNA is complementary in the frequency domain. We also noted that in DNA where spirals cross over, the vanishing is not gradual, since in these regions out of phase geometric phase, cannot cancel at wide ranges of frequencies. They disappear only at certain frequencies.

Our quantum cloaking differs significantly from earlier theoretical predictions. We have built our theory to explain the anomalous cloaking observed here. The theory predicts non-anomalous or naturally expected quantum cloaking for double helix, and prediction is experimentally verified in DNA. The fusion of electromagnetic lensing at resonance with the matching of the geometric phase path is limited to a helical path. However, this is very different from the established route of electromagnetic interaction with a quantum wave function. In the existing version, the potential or energy of an electromagnetic wave is linearly added to the wave function's potential or energy. That is a scaler addition of a value. Here, electromagnetic lensing needs a topology, and the topology of the evolution of geometric phase interacts to deliver the output of a quantum effect. Thus, it is not a mere addition but a match between two topologies, as a path of change in two kinds of symmetries.

3.9 Fourth Circuit Element: Hinductor

3.9.1 *The Memristor Proposal is Wrong: The Fourth Circuit Element Paradox*

There is a contradiction in the concept of the fourth circuit element proposed by Leon Chua. If we critically think about it, the memristor is not the true fourth circuit element. If the current flows, there is a linear change in the magnetic field, then the device works like an inductor. On the other hand, if the charge does not flow, generating magnetic flux without moving changes violates Maxwell's law. According to Maxwell's law, magnetic flux is produced by transmitting the current or physically moving charge. Here Maxwell's equations $\nabla \cdot E = \frac{\rho_v}{\varepsilon}$ (Gauss' law), $\nabla \cdot H = 0$ (Gauss's law for magnetism), $\nabla \times E = -\mu \cdot \frac{\partial H}{\partial t}$ (Faraday's law), $\nabla \times H = J + \varepsilon \cdot \frac{\partial E}{\partial t}$ (Ampere's law) are given. This paradox of inventing a fourth circuit element can be resolved by quantization of charge-flux. If the Dirac strings are in helical form, the magnetic monopole can coexist in ambient atmospheric conditions. To resolve the paradox, the idea of constructing such a usable particle device arises from the use of a genetic magnetic monopole generator (Fig. 3.7). To find a universe magnetic charge as a fundamental element like a magnetic unit (where the magnitude of the B vector is greater than that of the E vector), we do not need an accelerated charge as a result and an external massive magnetic field [81–84]. The additional benefit of such a magnetic element would be as follows.

3.9.2 Electric and Magnetic Field Vector (E&B) in Light

We do not consider memristor as the fourth circuit element since it cannot produce magnetic flux as a function of static stored charge. Instead, we have proposed a new kind of fourth circuit element called Hinductor that can store charge and produce magnetic flux. The structure we found for the true fourth circuit element consists of helical wires made of dots, known as quantum traps or wells. Such wells are designed so that the electric vector (E) is perpendicular to the surface of the well while the magnetic field (B) vector is parallel. The electric and magnetic vector contain a fraction of the charge density on the 2D surface of the virtual wall of the lenses. The magnetic field is increased one or two times by confining electrons and accelerating those by the depleted field through the topology of the surface, are well-known phenomena [85–87]. The depleted field can be efficiently changed on the topology of a surface by changing stored charges. Thus, variable phase correlation of coherent emission from such traps or well is similar to nano-antenna [88, 89]. In quantum and classical systems, coherent emission survives for a few picoseconds until the coherent resonance burst occurs. A spiral path is formed through phase-space interactions, while the optic axis confirms the interference of two waves (Figs. 3.11, 3.12, and 3.13). The magnetic flux then obtains a stable experimental standing wave

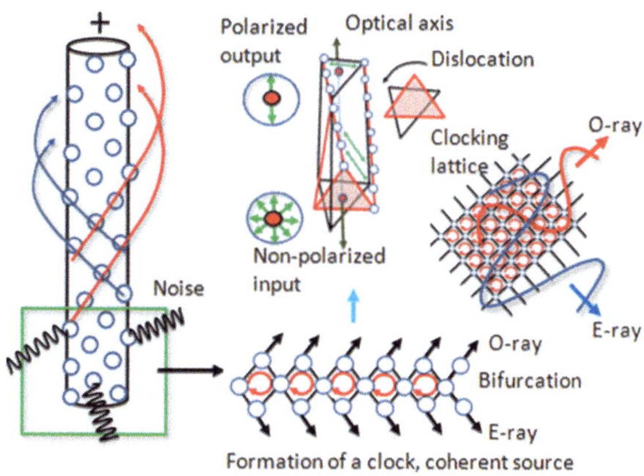

Fig. 3.11 All spiral-like structures or vortices have a direction of polarization that turns them into a signal used to organize a spiral's structure. Photons on the surface propagate by dislocation on a spiral and undergo E-rays and O-rays [81]. Two coherent sources together produce polarized light

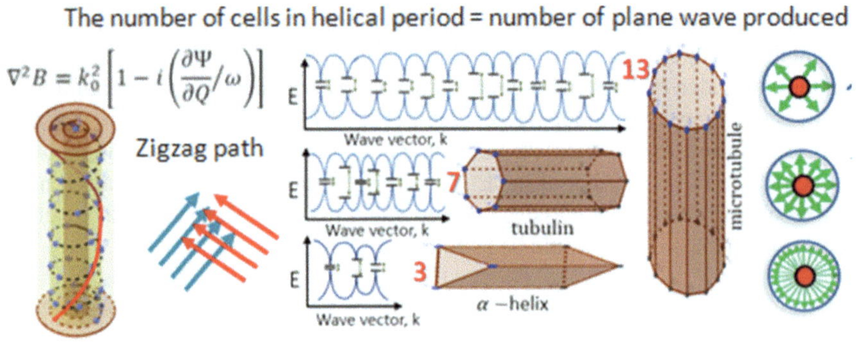

Fig. 3.12 This figure shows the spiral with 3 centers of interaction per cylinder, seven compositions of spirals or seven circuits for the tubulins, and finally, 13 photo filaments. The number of cells in a cross-section of cylinders or vortices denotes the number of plane-polarized signals produced by the device. The polarization charts of 3, 7, and 13 are shown [81]

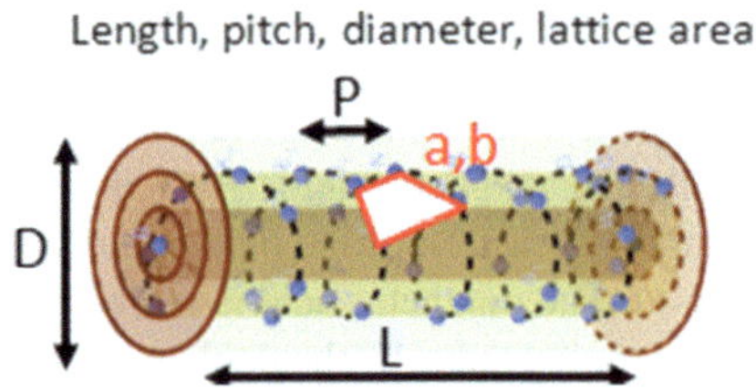

Fig. 3.13 Variation in the geometric parameters (length *L*, diameter *D*, pitch *P*, and lattice area **a** and **b**) of the device [81]

by arranging several wells in 3D form [90]. The interference on spiral pathways [91] can produce magnetic radiation different from the electrical role. A cluster or array structure helps to increase the magnetic field. Imaging of magnetic field is a complex task [92]. The magnetic vector interacts with the electrons of a sensor 10^4 times less than the electric vector, so imaging of the magnetic field in light is complex. Therefore, a special type of material metamaterial is used where the conduction of rings senses the magnetic ring faster [93].

The topologically conserved spin properties can be resonantly stimulated. Its elliptical dynamics have modes of clockwise and anti-clockwise that can be adjusted by pumping different MW (microwave) frequencies (Figs. 3.11 and 3.13). The transformation between two elliptical modes is obtained through a transition between different modes of linear fluctuations [94]. Therefore, phase modulation can be made by applying the THz frequency produced from the surface depletion field. Surface topology modulates infrared signals in the 5–6 THz frequency range into a 60–300 GHz lower frequency range signal. This is one of the most important down-conversion of the biological system that harvests thermal noise into a useful electromagnetic signal.

3.9.3 Theory of Fourth Circuit Element, Hinductor

Thus far, in diffraction, the topology of a surface is manipulated only to create static interference patterns. If the transverse component of an electromagnetic field is made zero by interference, then the wave turns to a moving line singularity, its direction goes indeterminant. If polarization is added to the field that perturbs interference, one could reshape the line singularity into a helix, a coiled-coil shape, even into a ring. The resultant 3D singularity structures could move like free particles. If a line singularity is perturbed, mostly it prefers to form a super-super coil. Therefore, if we begin with the concentric coil-triplets, all major geometric structures of line singularity could be created by adjusting the perturbation or coil geometry. We look for E singularity, H remains so that we get $E = B/C$.

Our proposed device is a coil made of balls of insulators and metallic cyclinder tubes (Fig. 3.14a). A ball is made of charge storage centers, which truly store the charge or increase the dwell time of charge passing through these centers. These centers are arranged in a cross-bar array. If the coil has a perimeter $2\pi r$, when it

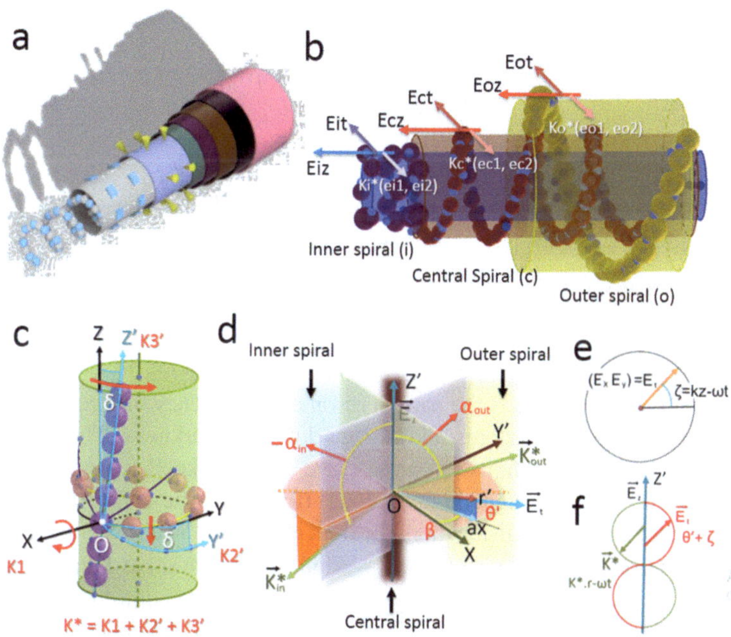

Fig. 3.14 **a** A photograph of a commercial H device. A helix antenna is connected to an array of coaxial atom probes and different metallic cylindrical tubes. **b** A schematic of the device with three concentric spiral antennas. **c** By twisting the spiral, it forms the screw and dislocation edges like microtubule and DNA. **d** A cross-section view of a schematic device is shown in panel b, and we have put the vectors in different directions. **e** Here, E_t is a transverse electric vector. Effective k^* neutralize E_t vectors. To get the magnetic vortex, the fourth circuit element-H device has to neutralize. **f** Magnetic vortices move like particles produced by the interference of layers

transforms into a spiral of length L, a spiral pitch p, a lattice forms with parameters a and b. The cylindrical surface has $\Omega = \frac{\text{Surface area}}{\text{lattice unit area}} = \frac{2\pi \text{rL}}{a \times b} = \frac{\pi \text{DL}}{\text{ab}}$, number of cells made of four reflecting corners. The wave that is flowing the charge clocks inside a cell. The zig-zag path sets the clocking frequency. A given spiral has several zig-zag paths; each path has a distinct cell. Depending on input noise, proper cells absorb quantized energy. If a pair of crossing paths have n_i number of clocking waves, each wave with an energy E_i, then, the total energy that is filtered from noise is $\sum_i E_i n_i^2$. The zig-zag path filters energy in a strictly quantized manner, the separated energy of the clocking wave is $\Delta\varepsilon_{\pm} = \frac{\hbar k_F^2}{2m^*} = \pm\sqrt{2(1-\cos k)}$, which is part of the total energy held by a pair of zig-zag path $K1$ and $K2$, is $\sum_i \frac{E_{K1-K2}}{2}\left((\cos\frac{\varphi}{\varphi_0})^2 + d_i^2(\sin\frac{\varphi}{\varphi_0})^2\right)^{1/2}$. Here, φ is the exchanged quantized flux between the paths (flux degeneracy) with respect to a flux minimum φ_0 and d_i is the physical separation between two parallel paths, k is the wave vector. The charge asymmetry between the paths holds an energy $\frac{1}{2}m^*\omega_L^2\left((1+\chi)x^2\right) + (1-\sigma)y^2 - \frac{\omega_{02}^2}{\omega_L^2}(z-z_0)^2)$, where χ is anisotropy factor, this anisotropy distinguishes between the diffracted waves from different paths. Here,

3D diffracted wave from a coil originates at the avoided crossing where the broken or fractured bands make a point contact. We get the 360° distribution as.

$$K = \frac{\Omega^3}{4\pi^2}\int_0^k (k_F^2 - k_r^2)N(P_z)dk \tag{3.1}$$

where $k_r = (1 - \frac{i\sigma}{\epsilon\omega})^{1/2}$, we get k_F^2 from the avoided crossing of the fractured bands, $\varepsilon_{\pm} = \frac{\hbar k_F^2}{2m^*} = \pm\sqrt{2(1-\cos k)}$, where the clocking charge oscillates between E_{F1} and E_{F2}. The conductivity of charges through the broken band surface $\sigma = \frac{nq^2}{m^*(\omega^2 - \upsilon_c^2)}(\upsilon_c - i\omega)$ and the probability of finding a charge between a pair of paths is $N(P_z) = \ln\left[\frac{1+\exp((E_{F1}-E_{F2})/\text{kT})}{1+\exp((E_{F1}-E_{F2}-\Delta)/\text{kT}}\right]$. Here, υ_c is the clocking frequency of a cell created by cross-bar trap, ω is the natural resonance frequency of a cell.

This 3D waveform interferes and creates knots of darkness. Thus far, it was believed that by suitable choice of surface, we could make $E = 0$, then, the dark lines should be made of magnetic field alone. If we twist the coil further, we get a mixture of edge and screw dislocation, the dark energy of the corresponding lines is given by $\frac{K^2}{2m^*}[\left(ax + kx^2\sin\delta\right) + i(y\cos\delta + z\sin\delta)]$. Here we solve K from Eq. (3.1), and the part $\left(ax + kx^2\sin\delta\right) + i(y\cos\delta + z\sin\delta)$ is solved by simplifying the defect on the coil surface, considering that a dark line is simply $\psi = k(ax + iy')e^{i(kz-\omega t)}$. We find three nested Bloch spheres that describe all possible static and dynamic dark lines from the derivation below. We simplify the 3D dynamics of such Bloch spheres, which do not have a classical point, but a fixed number of singularities. Along with X, Y, and Z directions, we get maximum $(x, y, z, xy, yz, zx) \times 2 = 12$ singularities, three dynamics of dark lines are pure along a particular direction, and four dynamics are paired effects. Thus, the phase function of a generic Bloch sphere that represents the singularity dynamics of our device is $\phi = 3\sum_i \cos\theta_i + 4\prod_i \cos\theta_i$, the wave vector of a dark line $k = \frac{\Omega^3}{4\pi^2}(3\sum_i \cos\theta_i + 4\prod_i \cos\theta_i)$. Since the wave vector on an anisotropic surface is $k = -\ln(hQ)$, h is the Plank constant, Q is the total charge distributed on the entire surface. The ratio of flux charge $H = \frac{\Delta\Psi}{\Delta Q} = \frac{he^{\Omega^3 k}}{hC_m} = e^{\Omega^3 k}/c_m$, here c_m is the velocity of light, hc_m is the action, the magnetic flux trapped in the smallest magnetic ring. Onc could see a brilliant blinking of the 12 holes in the phase space ϕ. For a typical classical device, we get an example of H, where P is the coil pitch,

$$H = \frac{1}{C_m} e^{\Omega^3\{3\cos(\frac{D-\mathrm{nb}}{b})+3\cos(\frac{P-\mathrm{na}}{a})+3\cos(\frac{L-\mathrm{nP}}{P})+4\cos(\frac{D-\mathrm{nb}}{b})\cos(\frac{P-\mathrm{na}}{a})\cos(\frac{L-\mathrm{nP}}{P})\}}$$

We soldered a large number of capacitors in a series (infinite resistance), and rolled it on a cylinder, varied diameter, pitch, and length, measured H, and imaged live the magnetic flux generated on the surface. The experimental values of H comprehensively match the above equation.

Including all factors here, we find the Hamiltonian **H** of the device.

$$\begin{aligned}\mathbf{H} = &\sum_i E_i n_i^2 + \frac{1}{2} m * \omega_L^2\left((1+\varepsilon)x^2\right) + (1-\sigma)y^2 - \frac{\omega_{02}^2}{\omega_L^2}(z - z_0))^2 \\ &+ \frac{\Omega^3}{4\pi^2}\left(\frac{2m*}{h^2}\right)^{\frac{3}{2}}\left[3\sum_i \cos\theta_i + 4\prod_i \cos\theta_i\right] \\ &+ \sum_i \frac{E_{K1-K2}}{2}\left(\left(\cos\frac{\varphi}{\varphi_0}\right)^2 + d_i^2\left(\sin\frac{\varphi}{\varphi_0}\right)^2\right)^{1/2} + \\ &+ \frac{K^2}{2m*}\left[\left(ax + kx^2\sin\delta\right) + i(y\cos\delta + z\sin\delta)\right)\end{aligned} \quad (3.2)$$

We built a device made of three concentric coils described in Fig. 3.14b to analyze the total effect of dark lines. The contact layer between a pair of coils is the YZ plane. Since coils are twisted, we get a mixed screw ($\delta = 0$; $\delta = \pi$; left-handed and right-handed respectively) & edge ($\delta = \pi/2$) dislocations on the coil surface. We realize this by rotating the X-axis by δ, so that YZ plane rotates by δ, ($x, y, z \to x, y', z'$; $k \to k^*, i.e. k_1, k_2, k_3 \to k_1, k_2', k_3'$) (Fig. 3.14c) We have shown that the spirals generate elliptically polarized plane wave k^* from noise. Each layer generates distinct waves, inner coil k_i^*, and outer coil k_o^*, (see Fig. 3.14d) which act as a perturbation to the central coil's straight stationary dark line k_c^* at an angle α and build singularity structures D_i and D_o, respectively. Inner and outer coils are internal and external mirrors, act as a pair of 180° out of phase sources, thus fit to edit perturbation phase 0°–180° and 180°–360°.

The dark lines ($E_t = (E_x, E_y) = 0$) form when the real parts of propagating signal $kr'e^{i(\theta'+\zeta)}$ and the two perturbation signals $e_1(e_1 e^{i(k^*\cdot r-\omega t)})$ and $e_2(e_2 e^{i(k^*\cdot r-\omega t+\Delta)})$ generated by the central coil's screw and edge dislocations cancel each other

(Fig. 3.14e, f). Different zig-zag paths generate distinct pairs of e_1 and e_2.

$$E_x = \mathrm{Re}\left\{kr'e^{i(\theta'+\zeta)} + e_1 e^{i(k^*\cdot r-\omega t)}\right\} = 0 \tag{3.3}$$

$$E_y = \mathrm{Re}\left\{ikr'e^{i(\theta'+\zeta)} + ie_2 e^{i(k^*\cdot r-\omega t+\Delta)}\right\} = 0 \tag{3.4}$$

We express e_1 and e_2 as $\overline{e} = (e_1+e_2)/2$ that represents total strength of perturbation and $\Delta e = (e_2-e_1)/2$ that represents polarization ($E_x \neq E_y$). Now, we consider two extreme cases, when $\overline{e} = 0$,

If two perturbations are complex conjugate, i.e. they are 180° out of phase, we get $\overline{e} = 0$.

$$E_x = \mathrm{Re}\left\{kr'e^{i(\theta'+\zeta)} - \Delta e e^{i(k^*\cdot r-\omega t)}\right\} = 0 \tag{3.5}$$

$$E_y = \mathrm{Re}\left\{ikr'e^{i(\theta'+\zeta)} + i\Delta e e^{i(k^*\cdot r-\omega t)}\right\} = 0 \tag{3.6}$$

(3.4) gives $\theta' + \zeta = -(k^*.r - \omega t) + 2n\pi$, where, $\zeta = kz - \omega t$. If we put k^* and r

$$\theta' + k_1 x + \left(k\sin\delta + k_2'\right)y' + \left(k\cos\delta + k_3'\right)z' + -2\omega t - 2n\pi = 0 \tag{3.7}$$

In both (3.3) and (3.4) we get, $kr' = \Delta e$, since (3.7) is a helix along Oz' axis, its cross-section is $r' = \Delta e/k$, and half-period π/ω. It means more is the polarization, wider is the helix. To get the helix pitch, we differentiate (3.7) and find

$$\Delta z' = \frac{2\pi}{k\cos\delta + k_3'}\{\frac{< 0\text{right} - \text{handed}}{> 0\text{left} - \text{handed}} \tag{3.8}$$

The helix moves with a velocity

$$v = 2\omega/(k\cos\delta + k_3') \tag{3.9}$$

Dark lines are straight lines along the z-axis, but perturbation transforms it into a spiral.

If $\Delta e = 0$, there is no polarization, in (3.5) and (3.6), $\overline{e}$ replaces Δe, but there is a change in sign in Eq. (3.5). Due to a change in sign, two things happen. First, the electromagnetic signal and the perturbation are not a complex conjugate as earlier, but inverse in sign, hence, ωt is canceled out from both sides, and instead of $2n\pi$ we add $(2n+1)\pi$ while equating the phase. If ωt disappears, it means the dark lines do not move; they are static structures.

Now (3.5) and (3.6) becomes.

$$E_x = \text{Re}\left\{kr'e^{i(\theta'+\zeta)} + \overline{e}e^{i(k^*\cdot r-\omega t)}\right\} = 0 \quad (3.10)$$

$$E_y = \text{Re}\left\{ikr'e^{i(\theta'+\zeta)} + i\overline{e}e^{i(k^*\cdot r-\omega t)}\right\} = 0 \quad (3.11)$$

By equating phase, we get.

$$\theta' - k_1x + \left(k\sin\delta - k_2'\right)y' + \left(k\cos\delta - k_3'\right)z' + (2n+1)\pi = 0 \quad (3.12)$$

This is a static helix with a radius $r' = \overline{e}/k$ and the pitch of the helix is.

$$\Delta z' = \frac{2\pi}{k\cos\delta - k_3'}\{\frac{< 0\,\text{right} - \text{handed}}{> 0\,\text{left} - \text{handed}} \quad (3.13)$$

Now, we can analyze the combined effect of $\overline{e}$ and Δe. When both are present, $\overline{e}$ would form a static helix and Δe try to form a dynamic helix. If $\Delta e < \overline{e}$, i.e., ($e_2 < 3e_1$) the static helix would be a guideline on which another dynamic helix would form, we get a supercoil. If additional perturbation arrives, we will get super-super coil; primarily we want three higher layers. The approximate solution we can derive for (3.14) and (3.15), when polarization is oscillatory $\Delta e = \sin\delta$, two spirals, made of magnetic null or electric null, twisting each other. Together, they form a cylinder $r' = \Delta e/k = \sin\delta/k = \lambda\sin\delta/2\pi$, the pitch of the helix and velocity are.

$$\Delta z' = \frac{\lambda\text{Sec}\delta}{2}\left\{\frac{< 0\text{right} - \text{handed}}{> 0\text{ left} - \text{handed}}\right. \text{ and velocity } v = c\text{Sec}\delta \quad (3.14)$$

We solve the generalized equations numerically to find the geometric details of line singularity.

$$E_x = \text{Re}\{kr'e^{i\left(\theta'+\zeta\right)} + \left(\overline{e} - \Delta e\right)e^{i(k^*.r-\omega t)}\} = 0 \quad (3.15)$$

$$E_y = \text{Re}\left\{ikr'e^{i\left(\theta'+\zeta\right)} + i(\overline{e} + \Delta e)e^{i(k^*.r-\omega t)}\right\} = 0 \quad (3.16)$$

Our numerical solution shows that the angle $\propto$ made by perturbation wave, i.e., elliptically polarized plane wave k^*, with the Oz′ axis classifies the singularity structures in two types. At $\propto < 0$, the average intensity $\overline{e}$ and polarization Δe both should be high (~2), to create a ring of electrical null, i.e., magnetic ring along with hairpin-like electrical singularity or magnetic structures. At $\propto > 0$, even if there is no polarization ($\Delta e = 0$), even a low-intensity signal ($\overline{e} \sim 1$) generates a ripple-like magnetic structure (no hairpin) in a twisted helix.

In the presence of additional perturbation from the inner and the outer coil, the produced geometric structures of vortices or fields are even more complex. When $\propto < 0$, the inner spiral generated signal is causing perturbation to the central coil signal. At $\propto > 0$, the outer coil generated signal causes perturbation. At $\propto > 0$, if there is no polarization ($\Delta e = 0$), then additional perturbation from the external coil generates a magnetic ring at even a low-intensity signal ($\overline{e} \sim 1$).

If $\propto$ switches sign, the clocking direction changes. If both inner and outer coil contributes together, we have a superposition of clockwise and anti-clockwise moving geometries. We use the term "superposition" because the perturbation is different, but two oppositely clocking geometries are produced from the same signal source in the central coil. If Δe varies periodically $\Delta e = \sin\delta$, then, $\propto$ switches sign periodically, we get a superposition of three patterns. Due to $\Delta e = \sin\delta$, the twisted spirals made of magnetic and electric dark lines are also created on the central coil. Infinite possible topologies could be created on the central coil for each of the three sets of patterns of dark lines; we get three interconnected Bloch spheres.

If we add more than three coils, an additional phase factor is considered by replacing e_2 with $e_2 e^{i\Delta}$, however, no new geometric shapes were found, additional coils are not profitable. In all E singularity structures, $E \neq 0$, but $E = B/C$ and two clocks make a donut or torus, i.e., free magnetic particle. The perturbation phase deforms the donut (a full circle ring) into a helical geometric phase, characteristic of a time crystal.

Numerical solution:

We simplify the above Eqs. (3.1) and (3.2)

$$\begin{aligned}
x &= \frac{r'}{a}\cos\theta' \\
y' &= r'\sin\theta' \\
s &= k\cdot r - \omega t = k_1 x + k_2' y' + k_3' z' - \omega t \\
\zeta &= kz - \omega t \\
z &= y'\sin\delta + z'\cos\delta \\
kr'\ \cos\left(\theta' + \zeta\right) &= -\, e_1\cos(s) \\
kr'\ \sin\left(\theta' + \zeta\right) &= -\, e_2\sin(s + \Delta)
\end{aligned}$$

For Hinductor, three concentric spirals are in contact via the yz plane, hence k^* lies in the yz plane making an angle α with Oz, positive toward Oy, we can simplify the above equations.

$$k_1 = 0, k_2^{'} = k\sin(\delta + \alpha), k_3^{'} = k\cos(\alpha + \delta)$$

and the identity we get $(kr^{'})^2 = (e_1\cos(s))^2 + (e_2\sin(s + \Delta))^2$, which gives $r^{'}$

In order to plot the dark line, we consider a function $D\left(s, \theta^{'}\right)$ to find zero level contour

$$D\left(s, \theta^{'}\right) = kr^{'}\cos\left(\theta^{'} + \zeta\right) + e_1\cos(s) = 0$$

Once we know s, θ' we convert $D(s, \theta') \rightarrow D(y', z')$, we find the cartesian coordinate of the interference null or knots of darkness.

3.10 Storage Charge and Magnetic Flux of H Device

3.10.1 Periodic Variations of the H Device: Theoretical and Experimental Validation

We have known from Fig. 3.15 that the variation in H is the function of resistance and impendence of a chemical solution CNT (Carbon Nano Tube). The Yagi antenna triggers the film of helical multiwall-CNT by applying the different frequencies. We can open or close the hole in the solution by changing the wavelength of light and composition of the solution [81, 82, 84]. We can analyze the relation of frequency (f), hinductance (H), and charge (Q) with the geometric parameters of the device from Figs. 3.15 and 3.16. The periodic oscillation of charge and flux is the main characteristic of the fourth circuit element. The maximum insulated cylinder provides a lower value of H.

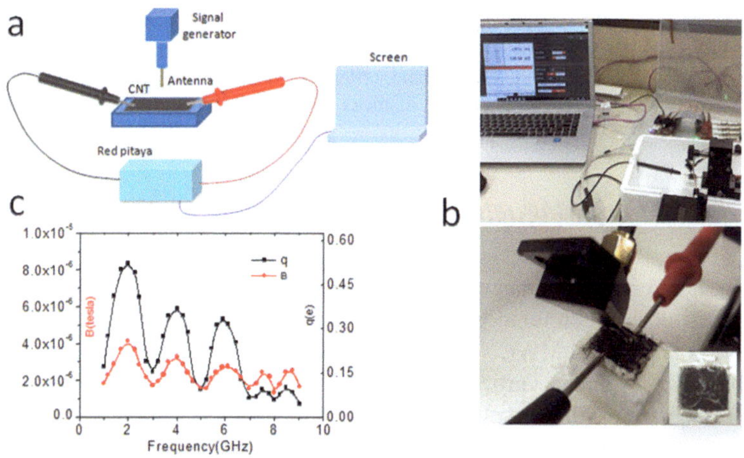

Fig. 3.15 **a** A schematic of the charge-flux measurement device. **b** An actual experimental setup to measure the charge-magnetic flux of a synthetic solution, MWCNT by applying the signals of different frequencies (1–10 GHz) through a Yagi antenna. **c** Representation of magnetic flux and charge (Y-axis) in terms of the frequency (X-axis)

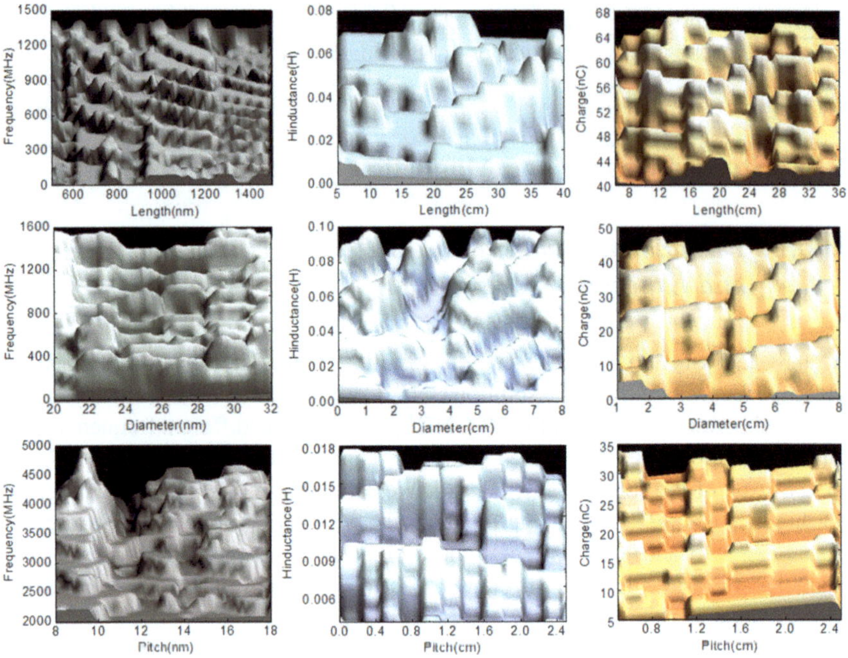

Fig. 3.16 The first column shows the variation in length, diameter, and pitch of theoretically built microtubules in terms of frequency. The second column is about the variation of H that relates to the geometrical parameters of the microtubule, observed experimentally. The variation in storage charge is the function of length, pitch, and diameter of artificial syntheses H device, as shown in column 3 [81]

3.10.2 Study of Magnetic Vortices

Here, we have created the magnetic vortices from various chemical solutions such as neurons extracted microtubule, multiwall carbon nanotube (MWCNT), nano brain solution, collagen, etc. MWCNT solution is triggered by applying frequencies from 1 to 10 GHz. Magnetic vortices pass through while optical vortices are blocked through the carbon film in solution. A magnetic vortex is a particle-like element. The magnetic vortex can propagate in the air or any non-magnetic medium after reflecting from the mirror (Fig. 3.17). We could change the optical vortex by shifting the position of the vortex lens, the synthesized solution, and the light source, but the magnetic vortex is always the same. The optical and magnetic vortices shown in Fig. 3.18, are derived by the solution of MWCNT. If we put the magnetic sensor at the position of an optical vortex, then we receive the signal without an electric vector ($E = B/c$) [81].

A basic setup of the magnetic vortices analyzer is shown in Fig. 3.19. It has a monochromatic light source (633 nm). The laser light falls on a lens (focal length 50 cm) and then passes through two polarizers. The polarized laser light passes

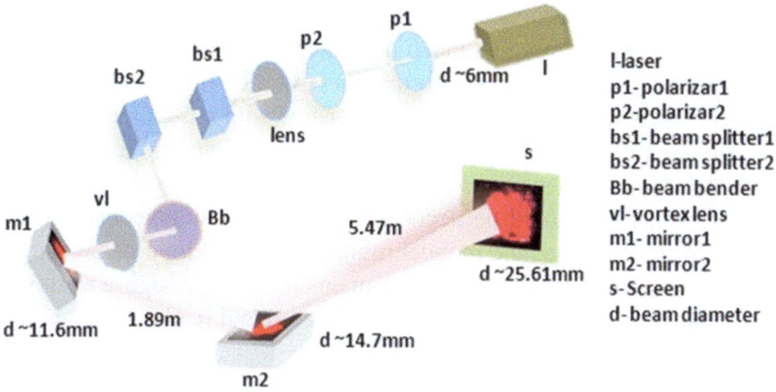

Fig. 3.17 A schematic diagram of magnetic vortex analyzer. The polarized laser light tilts by bs2, which causes it to focus on the beam bender (Bb). After beam bender, light passes through the vortex lens (vl), mirror (*m*1), and another distant mirror (*m*2), respectively. Finally, light falls on magnetic film and prints a shape of a magnetic vortex

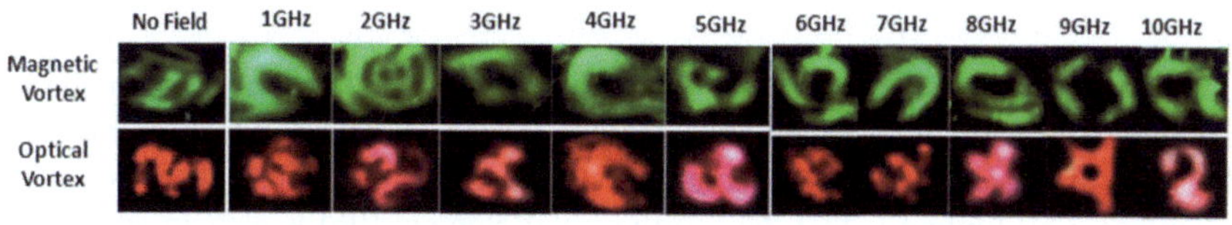

Fig. 3.18 Image of optical (bottom panel) and magnetic vortices (above panel) observed by pumping the different frequency signals (0–10 GHz) into MWCNT solution [81]

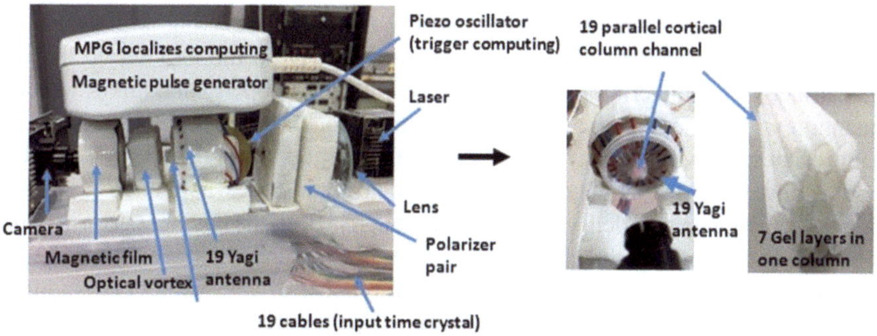

Fig. 3.19 A prototype of a big data analyzer [81], left panel. It consists of a laser, lens, a pair of polarizer, piezo oscillator, 19 Yagi antenna, magnetic pulse generator, optical vortex lens, magnetic film, and a high pixel camera. The right panel shows the different parts involved in the Yagi antenna channel such as the capillary tube in hexagonal form and 7 gel layers filled in it

through a channel of 19 cortical columns filled with a synthesized organic solution. The sample is triggered through a piezo oscillator, a magnetic pulse generator, and an array of 19 Yagi antennas. The information is written by 19 input time crystals; each time crystal is made of resonance frequencies with correlated phases. A time crystal is an assembly of clocks; it is fundamental to geometric musical language, GML [78–80]. Moreover, we have been modeling the biological systems using an assembly of clocks that could be decoded into many simultaneously propagating streams of waveforms. Once the monochromatic polarized light passes through the helical nanowires solution, it falls on the magnetic film through a vortex lens. A camera is located behind the magnetic film to record and read the printed magnetic vortices.

3.11 Conclusion: The Marriage Between Anomalous Quantum Cloaking and Fourth Circuit Element Hinductor

Long predicted quantum cloaking [95] is experimentally realized; a new, unpredicted phenomenon, anomalous quantum cloaking, is also observed, which extracts the dynamic of a part hidden in a complex structure that is impossible to visualize otherwise. Our finding is that the fourth circuit element [96] is also an anomalous quantum cloaking device. If we closely observe the theory, we would find that both the phenomena have most terms in the Hamiltonian common. Both phenomena are ingredients of brain jelly, organic material for new generation computing [97]. The physical reason for such a coexistence of Hinductor and cloaking is that both the devices require manipulation of surface charge density profile in a nested loop on the surface of the material. Evanescent wave holds the property and mixes with the quantum wave function for anomalous quantum cloaking, and when light falls, the same evanescent wave creates the vortices of electric and magnetic fields. That's why both Hinductor and cloaking devices have helical symmetry, where three concentric helices act in tandem to deliver the desired property. Both the devices perform better when we excite them with a noise burst, rather than a pure singular signal.

Quantum cloaking [95] promises to transmit matter as a wave, not bending light as frequently observed in classical cloaking. Using a wireless antenna, by pumping a suitable ac signal at resonance frequencies, we vanished a single DNA molecule repeatedly in its quantum tunneling images. Remarkably, apart from totally vanishing and reappearing at periodic frequencies, parts of DNA codes selectively appear and disappear. The same experiment on a single brain extracted microtubule reveals the dynamics of water channels deep inside it by vanishing the upper protein and ionic layers using inverse quantum cloaking. Surprisingly, the geometry of a helical coil of water regulates such anomalous quantum cloaking—paving the way to visualizing from far the atomic dynamics of a tiny part hidden inside a complex architecture.

References

1. Mitchell HH et al (1945) The chemical composition of the adult human body and its bearing on the biochemistry of growth. JBC158(3):625–637
2. Levy Y, Onuchic JN (2004) Water and proteins: a love-hate relationship. Proc Natl Acad Sci 101(10):3325–3326. https://doi.org/10.1073/pnas.0400157101
3. Sheinerman FB, Brooks CL (1998) Calculations on folding of segment B1 of streptococcal protein 278(2):439–445
4. Cheung MS, García AE, Onuchic JN (2002) Protein folding mediated by solvation: water expulsion and formation of the hydrophobic core occur after the structural collapse. PNAS 99(2):685–690. https://doi.org/10.1073/pnas.022387699
5. García AE, Onuchic JN (2016) Folding a protein in a computer: an atomic description of the folding/unfolding of protein A. PNAS 100(24):13898–13903. https://doi.org/10.1073/pnas.2335541100
6. Garegin A et al (2004) Water in protein structure prediction. PNAS 101(10):3352–3357. https://doi.org/10.1073/pnas.0307851100
7. Savage H, Wlodawer A (1986) Determination of water structure around biomolecules using X-ray and neutron diffraction methods. Methods Enzymol 127:162–183
8. Blundell TL, Johnson LN (1976) Protein crystallography. Academic Pr. ISBN 10:0121083500
9. Ernst JA et al (1995) Demonstration of positionally disordered water within a protein hydrophobic cavity by NMR. Science·267
10. Pal SK et al (2002) Biological water at the protein surface: dynamical solvation probed directly with femtosecond resolution. PNAS 99(4):1763–1768. https://doi.org/10.1073/pnas.042697899
11. Kauzmann W (1959) Some factors in the interpretation of protein denaturation. Adv Protein Chem 14:1–63
12. Raschke TM (2006) Water structure and interactions with protein surfaces. Curr Opin Struct Biol 16:152–159
13. Franzese G, Bianco V, Iskrov S (2010) Water at interface with protein. Food Biophys 6:186–198
14. Bellissent-Funel M-C et al (2016) Water determines the structure and dynamics of proteins. Chem Rev 116(13):7673–7697
15. Tarek M, Tobias DJ (2000) The dynamics of protein hydration water: a quantitative comparison of molecular dynamics simulations and neutron-scattering experiments 79(6):3244–3257
16. Ngai KL, Capaccioli S, Shinyashiki N (2008)The protein "glass" transition and the role of the solvent. J Phys Chem B 112(12):3826–3832. https://doi.org/10.1021/jp710462e
17. Martini S et al (2013) Water-protein interactions: the secret of protein dynamics. Sci World J 2013:6
18. Jasnin M, Stadler A, Tehei M, Zaccai G (2010) Specific cellular water dynamics observed in vivo by neutron scattering and NMR. Phys Chem Chem Phys 12(35):10154–10160
19. Takano K, Yamagata Y, Yutani K (2003) Buried water molecules contribute to the conformational stability of a protein. Protein Eng 16(1):5–9
20. Fisher S, Verma CS (1999) Binding of buried structural water increases the flexibility of proteins. Proc Natl Acad Sci USA 96(17):9613
21. Woessner DE (1977) Nuclear magnetic relaxation and structure in aqueous heterogenous systems. Mol Phys 34(4):899–920
22. Mathur-De VR (1980) The NMR studies of water in biological systems. Prog Biophys Mole Biol 35:103–134
23. Fuxreiter M, Mezei M, Simon I, Osman R (2005) Interfacial water as a "hydration fingerprint" in the noncognate complex of bamhi. Biophys J 89:903–911
24. Grossman M et al (2011) Correlated structural kinetics and retarded solvent dynamics at the metalloprotease active site. Nat Struct Mol Biol 18:1102–1108
25. Ansari S, Helms V (2005) Statistical analysis of predominantly transient protein–protein interfaces. Proteins 61:344–355

26. Berendsen HJC et al (1984) Molecular dynamics with coupling to an external bath. J Chem Phys 81:3684. https://doi.org/10.1063/1.448118
27. Schnepf MI (1992) Protein-water interactions. Biochem Food Proteins 1–33. https://doi.org/10.1007/978-1-4684-9895-0_1
28. Hanham SM et al (2015) Dielectric measurements of nanoliter liquids with a photonic crystal resonator at terahertz frequencies. Appl Phys Lett 107:032903. https://doi.org/10.1063/1.4927242
29. Chow E, Grot A, Mirkarimi LW, Sigalas M, Girolami G (2004) Ultracompact biochemical sensor built with two-dimensional photonic crystal microcavity. Opt Lett 29(10):1093–1095. https://doi.org/10.1364/OL.29.001093
30. Gabriel S et al (1996) The dielectric properties of biological tissues: III. Parametric models for the dielectric spectrum of tissues. Phys Med Biol 41:2271
31. Xu J, Plaxco KW, Allen SJ (2006) Probing the collective vibrational dynamics of a protein in liquid water by terahertz absorption spectroscopy. Protein Sci 15(5):1175–1181. https://doi.org/10.1110/ps.062073506
32. Bone S, Gascoyne PRC, Pethig R (1977) Dielectric properties of hydrated proteins at 9.9 GHz. J Chem Soc Faraday Trans 1: Phys Chem Condens Phases 73(0):1605. https://doi.org/10.1039/f19777301605
33. Damadian R (1971) Tumor detection by nuclear magnetic resonance. Science 171(3976):1151–1153. https://doi.org/10.1126/science.171.3976.1151
34. Inch WR et al (1974) Water content and proton spin relaxation time for neoplastic and non-neoplastic tissues from mice and humans. J Natl Cancer Inst 52(2):353–356. https://doi.org/10.1093/jnci/52.2.353
35. Webb SJ, Booth AD (1971) Microwave absorption by normal and tumor cells. Science 174(4004):72–74. https://doi.org/10.1126/science.174.4004.72
36. Pitera JW et al (2001) Dielectric properties of proteins from simulation: the effects of solvent, ligands, pH, and temperature 80(6):2546–2555
37. Li L et al (2013) On the dielectric "Constant" of proteins: smooth dielectric function for macromolecular modeling and its implementation in DelPhi. J Chem Theory Comput 9(4):2126–2136
38. Wyman J (1931) Bio Chem 90:444
39. Verma R, Daya KS, Understanding the decay of proteins: a method to study time dependent response of 5 pM concentration of insulin at microwave frequencies. https://doi.org/10.1016/j.mex.2016.11.004
40. Matyushov DV, Morozov AY (2011) Electrostatics of the protein-water interface and the dynamical transition in proteins. Phys Rev E 84:011908
41. Kyritsis A, Panagopoulou A, Pissis P, Serra RS, Ribelles JG, Shinyashiki N (2011)Water and protein dynamics in protein—Water mixtures over wide ranges of composition. In: 2011—14th international symposium on electrets, pp 49–50. https://doi.org/10.1109/ISE.2011.6084976
42. Ringe D, Petsko GA (2003) The 'glass transition' in protein dynamics: what it is, why it occurs, and how to exploit it. Biophys Chem 105(2003):667–680
43. Khodadadi S, Malkovskiy A, Kisliuk A, Sokolov AP (2010) A broad glass transition in hydrated proteins. Biochimica et Biophysica Acta 1804:15–19
44. Song M et al (2015) Wireless power transfer based on magnetic quadrupole coupling in dielectric resonators. Appl Phys Lett 108:2. https://doi.org/10.1063/1.4939789
45. Chabalko MJ, Sample AP (2014) Resonant cavity mode enabled wireless power transfer. Appl Phys Lett 105:24
46. Wang B et al (2011) Experiments on wireless power transfer with metamaterials. Appl Phys Lett 98:254101. https://doi.org/10.1063/1.3601927
47. Watts JL et al (2009) Fat synthesis and adiposity regulation in caenorhabditis elegans. Trends Endocrinol Metab 20(2):58–65
48. Ying L, Zhu H et al (2016) Current advances in the functional studies of fatty acids and fatty acid-derived lipids in C. Elegans Worm 5(3):e1184814

49. Hamilton LK et al (2015) Aberrant lipid metabolism in the forebrain niche suppresses adult neural stem cell proliferation in an animal model of Alzheimer's disease. Cell Stem Cell 17(4):397–411
50. Zhang SO et al (2010) Genetic and dietary regulation of lipid droplet expansion in Caenorhabditis elegans. PNAS 107(10):4640–4645. https://doi.org/10.1073/pnas.0912308107
51. Lemieux GA, Ashrafi K (2015) Neural regulatory pathways of feeding and fat in caenorhabditis elegans. Annu Rev Genet 49:13–38. https://doi.org/10.1146/annurev-genet-120213-092244
52. Bazinet RP, Layé S (2014) Polyunsaturated fatty acids and their metabolites in brain function and disease. Nat Rev Neurosci 15: 771–785
53. Vrablik TL, Watts JL (2013) Polyunsaturated fatty acid derived signaling in reproduction and development: insights from Caenorhabditis elegans and Drosophila melanogaster. Mol Reprod Dev 80:244–259. PMID:23440886. https://doi.org/10.1002/mrd.22167
54. Straub RH et al (2020) The memory of the fatty acid system. Progress Lipid Res 79:101049
55. Chang C-Y, Ke D-S, Chen J-Y (2009) Essential fatty acids and human brain. Acta Neurol Taiwan 18(4):231–241
56. Singh M et al (2005) Essential fatty acids, DHA and human brain. Indian J Pediatr 72(3):239–242
57. Das UN (2006) Essential fatty acids—A review. Curr Pharm Biotechnol 7(6):467–482
58. Fernstrom JD (1999) Effects of dietary polyunsaturated fatty acids on neuronal function. Lipids 34(2):161–169. https://doi.org/10.1007/s11745-999-0350-3
59. Simopoulos AP et al (2006) Evolutionary aspects of diet, the omega-6/omega-3 ratio and genetic variation: nutritional implications for chronic diseases. Biomed Pharmacother 60(2006):502–507
60. Yehuda S, Rabinovitz S, Mostofsky DI (2005) Essential fatty acids and the brain: from infancy to aging. Neurobiol Aging 26(Suppl 1):98–102
61. Salvati S et al (1996) Accelerated myelinogenesis by dietary lipids in rat brain. J Neurochem 67(4):1744–1750. https://doi.org/10.1046/j.1471-4159.1996.67041744.x
62. Salvati S et al (2000) Diet, lipids and brain development. Dev Neurosci 22(5–6):481–487. https://doi.org/10.1159/000017479
63. Stockard JE, Saste MD, Benford VJ, Barness L, Auestad N, Carver JD (2000) Effect of docosahexaenoic acid content of maternal diet on auditory brainstem conduction times in rat pups. Dev Neurosci 22(5–6):494–499. https://doi.org/10.1159/000017481
64. Kim JD et al (2019) Microglial UCP2 mediates inflammation and obesity induced by high-fat feeding. Cell Metab 30:952–962
65. Andrews ZB et al (2008) UCP2 mediates ghrelin's action on NPY/AgRP neurons by lowering free radicals. Nature 454:846–851
66. Schurig D et al (2006) Metamaterial electromagnetic cloak at microwave frequencies. Science 314:977–980
67. Tsang M, Psaltis D (2007) Theory of resonantly enhanced near-field imaging. Opt Express 15(19):11959–11970
68. Ammari H, Ciraolo G, Kang H, Lee H, Milton GW (2013) Anomalous localized resonance using a folded geometry in three dimensions. Proc R Soc A 469(2154):20130048
69. Baena JD et al (2005) Equivalent-circuit models for split-ring resonators and complementary split-ring resonators coupled to planar transmission lines. IEEE Trans Microw Theory Tech 53:1451–1460
70. Meklachi T, Milton GW, Onofrei D, Thaler AE, Funchess G (2016) Sensitivity of anomalous localized resonance phenomena with respect to dissipation. J Quart Appl Math 74:201–234
71. Nicorovici NA, McPhedran RC, Milton GW (1994) Optical and dielectric properties of partially resonant composites. Phys Rev B 49:8479
72. Saxena K, Singh P, Sahoo P, Sahu S, Ghosh S, Ray K, Fujita D, Bandyopadhyay A (2020) Fractal, scale free electromagnetic resonance of a single brain extracted microtubule nanowire, a single tubulin protein and a single neuron. Fractal Fract 4(2):11
73. Pendry JB (2000) Negative refraction makes a perfect lens (PDF). Phys Rev Lett 85(18):3966–3969

74. Jelinek L, Baena JD, Voves J, Marques R (2011) Metamaterial-inspired perfect tunnelling in semiconductor heterostructures. New J Phys 13:083011
75. Corbel S, Cerda J, Sautet P (1999) Ab initio calculations of scanning tunneling microscopy images within a scattering formalism. Phys Rev B 60:1989
76. Singh P et al (2021a) Quaternion, octonion to dodecanion manifold: stereographic projections from infinity lead to a self-operating mathematical universe. In: Singh P, Gupta RK, Ray K, Bandyopadhyay A (eds) Proceedings of international conference on trends in computational and cognitive engineering. Advances in intelligent systems and computing, vol 1169. Springer, Singapore. https://doi.org/10.1007/978-981-15-5414-8_5
77. Singh P et al (2021b) A space-time-topology-prime, stTS metric for a self-operating mathematical universe uses dodecanion geometric algebra of 2–20 D complex vectors. In: Ray K, Roy KC, Toshniwal SK, Sharma H, Bandyopadhyay A (eds) Proceedings of international conference on data science and applications. Lecture notes in networks and systems, vol 148. Springer, Singapore. https://doi.org/10.1007/978-981-15-7561-7_1
78. Bandyopadhyay A (2020a) Nanobrain: the making of an artificial brain from a time crystal. Taylor & Francis Inc. Imprint CRC Press Inc., Bosa Roca, United States, p 336. ISBN 10-1439875499. ISBN 13-9781439875490. https://doi.org/10.1201/9780429107771
79. Bandyopadhyay A, Ghosh S, Fujita D (2020b) Universal Geometric-musical language for big data processing in an assembly of clocking resonators, JP-2017-150171, 8/2/2017: World patent, WO 2019/026983; US Patent App. 16/635,900
80. Bandyopadhyay A, Ghosh S, Fujita D (2020c) Human brain like intelligent decision-making machine; JP-2017-150173; 8/2/2017; World patent WO 2019/026984; US Patent App. 16/635,892
81. Bandyopadhyay A (2020) Chapter 2 FIT, and GML; Chapter 3 PPM; Chapter 7. A complete, integrated time crystal model of a human brain. In: Nanobrain. The making of an artificial brain from a time crystal. Taylor & Francis Inc. Imprint CRC Press Inc., Publication City/Country Bosa Roca, United States, p 372
82. Reddy S et al (2018) A brain-like computer made of time crystal: could a metric of prime alone replace a user and alleviate programming forever? Stud Comput Intell 761:1–44
83. Sahu S, Fujita D, Bandyopadhyay A (2010) Inductor made of arrayed capacitors, Japanese patent has been issued on 20th August 2015 JP-511630 (world patent filed, this is the invention of fourth circuit element), US patent has been issued 9019685B2. Accessed 28th April 2015
84. Singh P et al (2020) A self-operating time crystal model of the human brain: can we replace entire brain hardware with a 3D fractal architecture of clocks alone? Information 11(5):238
85. Johnston MB et al (2000a) Theory of magnetic-field enhancement of surface-field terahertz emission. J Appl Phys 91:2104
86. Johnston MB et al (2000b) Simulation of terahertz generation at semiconductor surfaces. Phys Rev B 65:165301
87. Shan J et al (2001) Origin of magnetic field enhancement in the generation of terahertz radiation from semiconductor surfaces. Opt Lett 26:849
88. Biagioni P, Huang J-S, Hecht B (2012) Nano-antennas for visible and infrared radiation. Rep Prog Phys 75:024402
89. Michler P et al (2000) Quantum correlation among photons from a single quantum dot at room temperature. Nature 406:968–970
90. Empedocles SA, Neuhauser R, Bawendi MG (1999) Three-dimensional orientation measurements of symmetric single chromophores using polarization microscopy. Nature 399:126–130
91. Freed S, Weissman SI (1941) Multiple nature of elementary sources of radiation—Wide-angle interference. Phys Rev 60:440
92. Vignolini S et al (2010) Magnetic imaging in photonic crystal micro cavities. Phys Rev Lett 105:123902
93. Schirber M (2010) Measuring the magnetism of light. Phys Rev Focus 26:13
94. Jin C et al (2017) Topological trajectories of a magnetic skyrmion with an in-plane microwave magnetic field featured. J Appl Phys 122:223901

95. Bandyopadhyay A, Fujita D (2021a) Method for realizing quantum cloaking in electromagnetic device for remote imaging apparatus. Application no. 2021-172701
96. Bandyopadhyay A, Fujita D (2021b) Electromagnetic device, magnetic and electrical vortex synthesis device and magnetic and optical vortex synthesis device. Application no. 2021-172702
97. Bandyopadhyay A, Sahoo P, Fujita D (2021c) Self-learning by information processing device and self-learning for information processing method. Application no. 2021-172703

Chapter 4
How to Reverse Engineer an Organic Human Brain Without Using Any Chemicals?

4.1 Brain is not a Chemical Soup: It is a Massive Clock Architecture

4.1.1 Five Radical Changes in the Concept of Biology

4.1.1.1 Changing the Way We Think: Ions and Chemicals Could Be Millisecond Clocks

Current biology books teach us that the brain is a chemical device, and almost all key biological information processing happens chemically. The current premises of biological information processing argue that molecules and ions carry the information. The chemical state is the information, and the time taken for chemical interaction has no importance, only the chemical state matters. We are advocating an alternate view. Our view is that we should not get overwhelmed by the complex chemical processes but look only into the time domains used by biomaterials. There are structures of various dimensions, one inside another. We have listed at least 12 layers of structures are there from the molecular scale to the largest. Twelve different carriers, solitons like defect state, electrons to electromagnetic waves or photons, ions, molecules, electrical pulses, are abundantly found in a single biosystem. Looking into the time domain means we can make a time scale and forget their structures' complexity and wide variations. The spatial arrangement of clocks could alternately model the biosystems and that could change biology forever. Imagine, for every single event from the atomic scale to the meter scale, 10^6 orders of spatial scale variation are mapped by, say, 10^6 orders of temporal scale. There would be plenty of advantages of changing the worldview. In order to process milliseconds signals, biological systems process ions, and molecules. Because ions electromagnetically resonate in kHz or waves that change amplitude 1000 times per second, equivalent to milliseconds operations. Here we outline some advantages.

P. Singh et al., *Biological Antenna to the Humanoid Bot*, Studies in Rhythm Engineering,
https://doi.org/10.1007/978-981-16-9677-0_4

1. **Integrate all constituent and hierarchically correlated functional systems into one model**: Electromagnetic resonance chains [1] would integrate all biologically connected materials along with their function. Our brain is a linguistic machine where the information element is the triplet of time crystals [2], which can be considered useful for understanding the language of nature [3].
2. **Noise-like signal bursts become important because they hold geometric shapes: water becomes important because it couples the geometric shapes or noise bursts**: Each signal is converted into a pulse by our brain. Inconsistency in the shape of the pulse determines the language of the brain. Information is in our brain as geometric [4]. The combined fluctuation of the peaks and valleys of information generates human thoughts [5], then the fluctuation in time is neuroscience [6]. When we change the worldview, the shape of the signal burst becomes important because it holds a geometric shape. Consequently, the water becomes important as it interfaces two vibrating components integrated into a biological system.
3. **Biological rhythm is redefined: A universal clock architecture that integrates all forms of rhythms**: The initial fundamental law of rhythm forms a certain class of rhythms distinctly recognized by their sensory sources. The second law is the superposition of various chains of rhythm. The third law is the pairing and de-pairing between different signals coming from different sensory organs.
4. **Suddenly the vibrating strings located deep inside a neuron turn active: Century-old faith that membrane does everything is challenged**: Currently, neuroscience considers that neurons are everything in the entire brain. Everything is silent within neurons. The membrane of a neuron does everything, and at the neuron level, wiring transmits the information in the brain [7, 8]. The chain of neurons is important in the brain mapping projects, although they abstain from the idea that those neurons use clocks for conversion [9].
5. **An architecture of symmetries with no physical attribute takes over as the supreme controller of biosystems**: As two human brains do not have the same route and new links develop every moment. Thus it requires a basic type of verbal communication to examine the pathway of cognition that does not depend on the particular type of wires. One such constraint is (Phase Prime Metric-Geometric Musical Language-Hinductor) PPM-GML-H triad; [7, 10, 11] the Intelligence engages within and on top of the membrane of a neuron. The arrival of dodecanion tensors and their maniflats (not manifolds) opens the door to a set of primes representing thoughts and emotions. It is a revolutionary change in developing a self-operating mathematical universe (SOMU).

4.1.1.2 The New Kind of Brain Circuits Made of Time Crystals

Thus far, the brain was considered a linear circuit because it has been a firm belief that the Turing machine is the ultimate computing device in the universe. It can model any information structure that nature can produce. Consequently, the brain is a circuit, and we understand its governing principle. What that all we need is resources, and then making a brain would not be difficult. We reject this view based on our biophysical studies described in the first three chapters of this book. Here is a summary of what we have learned in our biophysics study.

1. **Poly-atomic time crystal engineering**: The 3D spatial arrangement of clocks is the information structure of a biological system. To operate clocks, biological systems use different carriers like ions, molecules, charges, photons, defects of various nature. Biological structures are designed to synthesize clocks, and they self-assemble to integrate clocks. We call it polyatomic time crystal engineering. It is about crystals made of different atoms or clocks.
2. **Noise burst in geometric shape is the true nature of signal used**: Since biological systems mostly operate using a high density of clocks, the signals emitted by biological systems include compositions of many frequencies. Therefore they mostly appear like noise bursts. The burst shape is similar to the geometric shape that biomaterials encode. Therefore noise is the mode of signal transmission if we use conventional instruments to read them.
3. **Water filter for splitting coherent signals and thus assist in filtering noise**: Three layers of water molecules cover the biomaterial structures. They act as integral signal processing elements for the biosystem; they also act as filters to geometric shape noise and convert them into clocks. Water is, therefore, an integral part of biological information processing.
4. **Anomalous quantum cloaking**: Since biological components primarily do not connect with physical nerve fiber wiring unless it is essential to send a specific signal to a specific component alone, wireless communication is fundamental to biological systems. Biomaterials make leaky cavity resonators and dielectric resonators. Selective invisibility or anomalous cloaking is a key to such communication [12].
5. **Synthesis of the vortex and vortex condensate from noise**: Vortices are ring-shaped fields or flow of carriers. A typical geometric shape ensures stability, and the shape acts as a virtual atom. Several vortices could condense into a 3D architecture and act as a molecule, supramolecule, etc. Biological structures are designed to generate vortices even by splitting photons into electric and magnetic fields [13]. Fundamental engineering required to build a biomaterial is harvesting photons for building a loop of fields on its surface and encoding the loop properties in the evanescent bursts.

4.1.1.3 Resonance Chain Synthesizing Environment as a Polyatomic Time Crystal

When the external environment changes, the ripple effect comes into our brain through a particular frequency domain of the brain mapping diagram. However, a pulse of light can reset the brain. When we listen to someone, language produces a map of time [14] within the brain. Language is like a tune or music [15, 16]. The smoothness of a brain is active in the space [17]. The resonance in the hemisphere of ions affects the human brain rhythms [18]. The temporal disorder is the key to solving several mental disorders [19]. When a rhythm of a large period begins to change, the brain and the entire body undergo the smallest precise action in different membranes to regulate the modified beats, which we know as the decision of the human brain. A resonant chain of whole-body connects of all brain components as a triplet of triplet rhythms [2, 20], and it synchronizes with the perturbation of external environment to the internal rhythms. However, researchers link the functional areas of the brain. A study on event-based coherent communication of brain elements is reported [21].

One of the primary decision-making domains in the brain is the cortex area. The cortex region is made of hexagonal close packing of cortical columns. The function of the cortical column could be presented using 3D assembly of clocks and similarly we could build 3D clock assembly for the protein circuits. However, both the systems build clocks in different ranges of frequency. Therefore, we prepared several maps at a different scale of time for different elements of the brain. All these charts will help in understanding the information in the brain. Information in the brain is a noise burst that has a geometric shape. These geometric letters of the brain's map vibrate or form a series of vibrations forming time crystals that fix or bond the circuits of all layers [7]. The formations of such a diagram would allow one to learn scientific behaviors applying hierarchical sequential maps of the brain.

4.1.2 Information Propagation in the Biological System: All Elements of the System Contribute Equally to Information Processing

4.1.2.1 Brain as a Spatio-Temporal Fractal Machine

The neuron is an element to synthesize the brain's power [22]. Many layers are found in the brain-body system, one inside another. Circadian pulses rotate in every layer [23, 24]. Not only a particular layer can be the root of information but all layers from the architecture of the time, which are the source of information. The state variation between two cells cannot be described since many cells are inside it, and outside of each cell, there are also many cells. So the information state of a cell is not complete. As stated, no element touches any other element in a fractal network, and each unit is the entrance to a new self-similar structure. What device will work here? We have proposed the fourth circuit element, Hinductor, in Chap. 3, which was inspired by the microtubule design. Therefore, the resonance chain where each resonating unit is a Hinductor or H device is a simple way to communicate with the entire system. This suggestion is better than the old suggestion; 'Brain circulates the information using time' [25–27]. The new paradigm would be that the brain's information is the symmetry of time and prime with no physical attribute.

4.1.2.2 Brain is a Machine that Only Rearranges Primes

Previous brain studies on rhythms were based on only ionic rhythms [28]. Now we have considered all possible states above and inside the neuron level. The electromagnetic idea-based brain was proposed long ago [29]. The purpose of creating the brain will be such that the twelve elements of the brain will process through limited numbers of time crystals using 15 primes [7].

4.2 The Quest for Primes in the Brain Engineering

4.2.1 Generic Device of Decision-Making of Human Brain

The brain cannot be a discrete sum of Turing machines because it is made of nested clocks grown within and above. There could not be a single defined state because every clock leads to an infinite chain of clocks [7]. The neuro-magnetic rhythm is circulated in the human brain [30]. Each brain element vibrates with frequencies whose ratios are a suitable set of prime numbers. Consequently, the resonance frequency band of a brain component could also be represented by a pattern of primes. The phase prime metric made of ordered factor is formed from the resonance series. To search the prime in the resonance band of brain elements (Fig. 4.1), we have to look at the brain's architecture. First, we find the primes in the electromagnetic resonance band. The symmetry of primes is evident in the structural symmetry. A higher dimensional complex tensor is created to address hierarchical interactions between

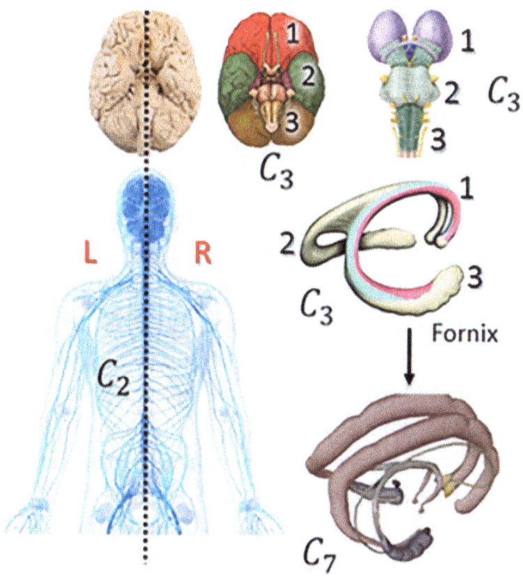

Fig. 4.1 The neural system of the brain follows C2 symmetry (left panel). Three major lobes of the brain show C3 symmetry (mid panel). Similarly, C3 symmetry is found in the brain stem. In the midbrain region, the fornix and cingulate gyrus have C3 and C7 symmetry, respectively (right panel) [7]

primes. This tensor is the key to brain information in the form of time crystals [8, 31, 32]. So, what does this little chain do?

1. The little chain formed by a guest brain component vibrates with the large chain of the host biological component to exchange the time crystal.
2. The rhythms will expand within all of them from which they will self-assemble.
3. During phase conversion, some rhythms become weak, and some become strong.
4. Reply of time crystal to each other quickly by the magnetically, electrically, and mechanically resonance.
5. Harvest the noise and filter it into a signal [33] for introducing a new AI [34]. Since the current AI is based on the concept that neuron is everything, and their membrane is responsible for information processing.

However, in our case, the whole body system is the brain. We continuously go inside at the molecular level to the resonance chain of the devices. The device does not end up within our body, and it receives matric from the environment. In a cellular system, the principle of design of the synchronization oscillator obtained by nature shows the geometric language we use to recognize patterns. The little oscillators are organized in initial geometric shapes to make all possible oscillators. Different symmetries are found in the arrangement of the human body and its elements, for example, whole human body-C2, Fornix and cingulated gyrus-C3 and C7 respectively (Fig. 4.1), etc. [35]. The time difference between the interhemispheric of the brain shows how the information concentrates on the left and right brain or has the C2 symmetry.

4.2.2 *Significance of Time Crystal Map of the Human Brain*

The temporal architecture of the human brain has attracted the attention of many researchers [36]. There are different kinds of links and connectivity within the human brain, which keep changing regularly. Mechanics is not well-known to an observer [37], which is the primary stage to understanding the brain.

Natural vibrations are the property of every material. The rhythms of atoms at a large scale can be built into a complete map of the brain that holds some special features which are not limited to building the hardware of the human brain. Species must have the same rhythm; otherwise, they cannot interact with each other [38].

All phenomena can be understood from the case of symmetry breaking, whether it is related to quantum or classical mechanical or any chemical or physical activity at any level. Therefore, we can identify every event from the case of symmetry breaking. The divergence of the brain size across species exhibits wiring differently along the brain to maintain symmetry [39].

A range of rhythms incorporates nature in its cycle at a fixed scale. To send even an image, the neuron is formed the structure and transmission of time [40]. This is the reason why our body learns and fits with the environment without any cognitive effort. The organisms act as a computer; they survive, learn, and produce (birth) a new computer and finish.

As we have suggested, the brain is not a computer. Universal time crystal claims that there is no communication [41]. When our entire body is a part of the time crystal composite, there is no requirement to run distinct signal transmission.

Many elements in the brain are fractal-like, self-similar, and of similar symmetry. They sometimes change their order as a result of which their symmetry is broken. When a system becomes large, it does not matter how much symmetry its components have. The larger system also follows only a small number of symmetries. Reddy et al. [42] showed a matrix of primes in the brain that start from one, accounting for the initial fifteen primes or stops at forty-seven.

Symmetry is the primary, and shape is a secondary key of a topology. Therefore, the number of neurons or filament nanowires does not matter [43]. Symmetry is the means of exchanging language from each other. Without it, we cannot identify the language, and we aim to find it.

The universe is multi-dimensional and largely filled with basic elements. While a time crystal suggests a reset curve of phase and a composition of phase reset curves suggests a composition of a time crystal, believed for the brain [44]. It would be interesting to see how the brain map defines the universe. Singularities in a clock are key to a time crystal structure [45]. The phase diagram of a group of oscillators shows that there is a combined and undefined phase activity. The human brain probably integrates multiple clocks [46]. Integrated clocks depend on the nature of the clocks' behavior and the groups of clocks of different behavior that are phase-locked with each other [47]. The rhythm of the motor suggests a 3D geometric model of clocks [48].

The conversion from rate code to temporal code has been studied for a long time [49]. This conversion is generally present in the standard time crystal model of the human brain. The time crystal model considers the rules of sub-threshold signals that perform complex conversions of signals in neurons [50] but does not take them seriously in brain models. The link between the rate code and the sub-threshold signal may show a new information processing path in the brain [51].

4.2.3 Fractal Behavior of Resonance Frequency in Components of the Human Brain

The structural form of each element of the brain is available online. Resonance bands can be found by resolving the brain elements in EM simulation software. It is possible to print an element through a 3D printer, detect its resonance band, and verify it. Some conventional brain copies are attempting reverse engineering of the brain [52] by considering a neuron's skin or membrane pathway. Elements of the brain vibrate using a singular natural logarithm [53]. Here, we have shown our work by the clocking properties of the entire brain-body system, not omitting any component. One has to plot the transmission and reflection of the components of the brain as the resonance peaks in terms of frequency. If we take the log of frequency on the principal axis, then the design (plot) looks as if the log scale separates the resonant frequencies.

Whereas, after taking the log scale of the peaks, they should appear equally spaced. The distribution of frequencies is the log scale inside the log scale. To remove log distribution, the values of the log are separated by a linear distance. If we consider the linear values and plot the derivative of its resonance frequency, we obtain the log distribution again. The log characteristic or non-linearity cannot be reduced [7]. As a result, we can separate the frequencies by the log function inside the log function. Probably, this property is found in both the resonance chain and the brain. This is a nested frequency fractal. A condition of making a time crystal is non-differentiable. Suppose the multilayered seed structure has an equal distribution of power at the value of each resonance frequency when it is constructed. In that case, the structure must accept a symmetry that allows for continued same power loss overall structure. If the same power loss is preserved in a scale-free approach, lower frequencies must be near more. Now the law of power is a conservative argument or claim. An exponent of a power relationship holds an endless chain. Therefore, a log inside a log inside a log… (not as log (log (log (frequency….))) is not differential, so it is non-Turing.

4.2.4 Resonance Band Featuring in Terms of a Triplet of Triplets: Universal Resonance Chain

There are rhythms of different nature in the brain. The waves of heat flow interact with the wave of ions. Here, all resonances are interlinked. The phase prime metric (Fig. 4.2) is the triplet coupling of the distribution of the divisor choices of the integers [7]. The triplet pattern is found in the shape of the protein complex, such as the microtubule. Even the peaks in the resonance spectrum of the human brain structure revealed by the experiment also show the triplet pattern. The plot of universal time crystal looks teardrop or pear-shaped. The parametric equation of the resonance

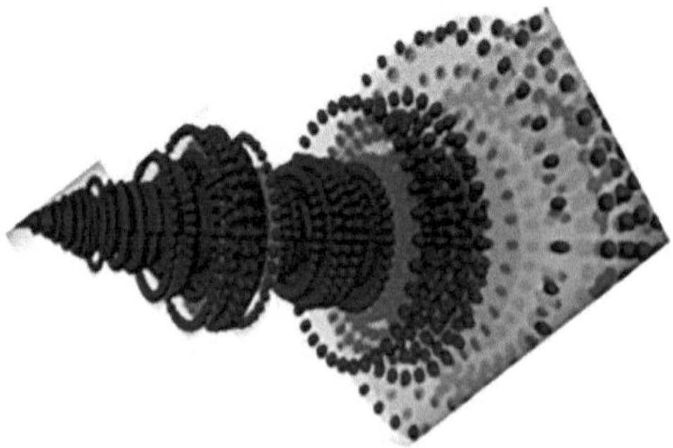

Fig. 4.2 A photograph of phase prime metric, PPM, is shown here [7]. Each ball represents an event. PPM means all possible combinations of a given number of events that can be grouped

chain for the curve of teardrops will be $x = \cos t$, $y = \sin t \sin^m(t/2)$ and for the curve of the pear-shaped, we find $b^2y^2 = x^3(a - x)$. The Inverted Mandelbrot design is also in the shape of a droplet, but here we have not noticed the shape but the hidden pattern. Form the universal resonance chain; one can produce the basic constants. As the brain has the character of the universe, geometric constraints are also similar to the universe.

4.3 Foundation of a Clock in Biological Organism

The vibrations of matter are cyclic in which matter moves in a circular form, and it is necessary to walk in a loop to connect the gap. That vibration arises from the material that exchanges energy between two participating systems. When exchanging energy, a part of that energy is not found in any participating system, which we call bond energy. It can occur at any level. When a matter receives a packet of energy, this energy goes to structural symmetry. Symmetry always takes energy to vibrate such parts, due to which it has special vibrational frequencies. Areas of similar symmetry start vibrating when they find the energy packages [8]. Energy can be exchanged by sending photons between two matters when the medium is pumped between the sending materials. All of the three substances, or a pair or individually, can exchange energy. If energy is exchanged once in a matter, they are not coupled. If there are one or more times, the energy exchange starts cyclically, and cyclic rhythms are produced. Temporal organization in spatially arranged neuronal collections shows a hierarchical network in clocks [54].

4.3.1 Meander Flower Garden

Meander flower is a special contribution to the study of the time crystal (1970–2000; Fig. 4.3a–c). Meander flowers are similar to classes of epicycloids, hypocycloids, and cycloids that can be identified by placing the guest clock at different locations of the host clock. These geometries were described as a planer curve more than a decade ago [55]. Nested clocks are the function of time [56]. When someone observes the EM resonance band of the biomaterials, the resonance peaks form groups. These groups are the main features of the geometric shapes. To understand, if three peaks oscillate in a group, there is a triangle encoded. How these frequency groups change through additional input signals such as electromagnetic, magnetic, mechanical, electromechanical, or ionic vortices is an important question of the brain. The compositions of vortex atoms form time crystals and conduct unique geometric such as geometrical musical language (GML) and phase-prime metrics (PPM) [42]. Different PPMs work simultaneously for each type of vortex atom. Each PPM is a meander flower garden, and the combination of the PPM is the Garden of Gardens

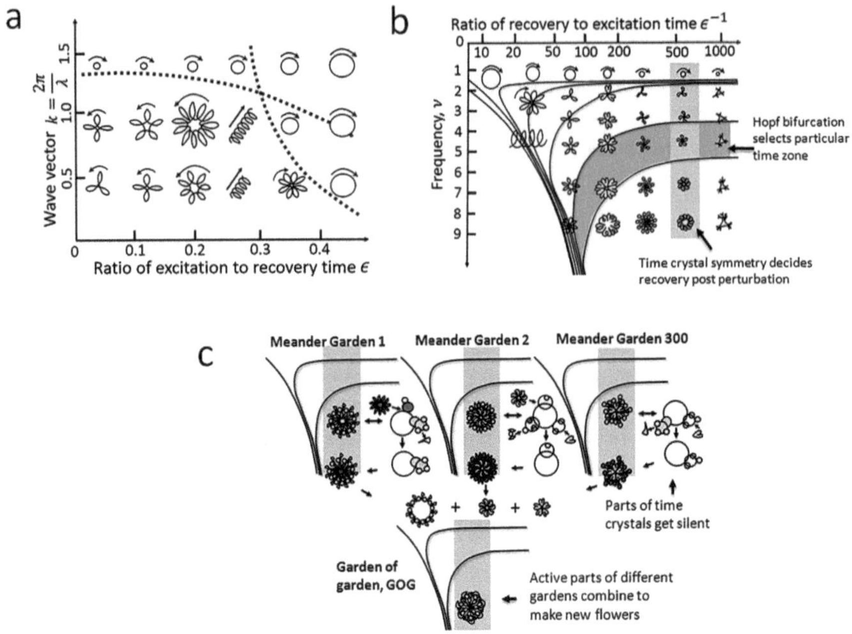

Fig. 4.3 **a** The recovery time (the time to return to the previous shape) depends on the input perturbation time and frequency of the signal used. Therefore, nested clocks are different in terms of phase. By changing the diameter, we can make many time crystals. Every time crystal has two clocks, signal, and system point, we find the meander flowers, and combinations of meander flowers are meander flower garden. **b** The design of Meander Flower Garden, similar to panel (**a**) is shown, but the number of waves is changed by frequency. Here the ratio of excitation and recovery time is inverse. The shaded parts show the two main columns of the sequential generation of time crystals. **c** Many gardens of meander flowers can interact and create the garden of gardens where several meander flowers combine and make a new flower [8]

(Fig. 4.3a–c; [7]). Even though an image contains many objects, the brain has to identify abstract geometric relationships between objects.

When different elements of artificial brain hardware receive sensory signals of multiple classes, the patterns present in the elements correlate and form a fractal seed structure due to the natural character of the frequency fractal hardware. The circuit of the brain undergoes minute changes to integrate these features. In this way, data or information of touch, visual, taste, sound, and smell is correlated in the brain hardware. Such data should be considered the highest operation of the brain's fractal hardware, known as Brodmann areas of 47 cortex sections. New input fractals drawn from geometric shapes are stored in the brain only when it is not identical to the patterns available on the elements. If it is the same, there is no need to store everything new for the hardware. Placing a flower in the plot is an idea, and incorporating a garden is the learning potential.

4.3.2 Transition from Meander Flower to Meander Garden by Time Crystal

The rhythms of music have explained the structure of the brain-body. For example, thoughts in the brain [57] are often responsible for cerebral rhythm [58]. Different elements of the brain react to the rhythms of music in different ways. In particular, musical rhythms [59] can be felt in the EEG of the brain. Music-like rhythms are produced during motor learning [60]. The states of the elements present in the brain are oscillating. If the information in the brain is treated as a geometry, then the state of the brain can be clarified by neural oscillations [4]. From the diagram of temporal disorders, it is known that a group of clocks can move from one structure to another, and it is possible to build a new structure [19]. This mechanism is the basis of the decision-making process of the brain. A brain consists of a group of composite time crystals; from these groups, the brain selects small clusters of time crystals and builds unprecedented time crystals. The garden of the garden selects the flowers from different gardens, and they synchronize to create new flowers [61]. The transition from clocks to chaos (disorder), chaos to clocks, occurs continuously [62]. The absence of the garden of gardens is seen as unconscious, yet entire gardens remain active [63].

4.4 Different Perspective from Existing Neuroscience

4.4.1 Operational Language of a New Time Crystal

Phase singularity in time crystals is always a reaction to an external perturbation. Vortex and spiral-shaped devices create time crystals. Phase singularity is attached to the geometry of the device. It is fundamental. It helps us to use singularly to develop the engineering of self-assembly, noise reduction, information processing, etc. The language of geometric is operated by the blinking of singularity or holes in the phase space of the spiral, where all incoming information is transformed into a 3D shape. In geometric musical language, information is in the form of a group of different geometric shapes, which have singularity points at the corner where the energy bursts. A system point rotates in a cyclic loop and touches singularity points that create a burst of energy. So, the total number of burst points equals the total number of corners in the geometry. The information between burst points is in a loop that counts the ratio of the primes, which is the symmetry of the geometric shape. The pattern PPM (phase prime matrix) created from the order factor of the primes combines all symmetric events in the universe as geometric shapes.

4.4.2 Singularity: Origin of a Clock

The oscillations of high-frequency occur at 1 kHz frequency. Such oscillations remain at a limited resolution of the time of ionic impulses [64]. One of the difficulties in explaining ionic transmission at high-frequency is that if the time of rhythm reaches the limit of ionic transmission and cognition occurs within that time range, control must occur at a faster time scale [65]. Although ionic transmission has attained its limits, we need another way of communication. In current neuroscience, there are no other ways of communication. The geometric language that describes singularity can fulfill its need. The conversion from one time to another occurs through a burst from the singularity point (Fig. 4.4). A clock can be shown by a maximum of four singularity points using a 2 × 2 tensor [31, 32]. In Fig. 4.4, a diagonal is drawn to show the C2 symmetry of the clock and the structure that makes the clock. The main property of the brain's architecture is a one-to-one correspondence between the shape and its symmetry. The geometric shape of information is forever unseen.

All sensors (ears, eyes, skin, tongue, and nose) use a dodecanian tensor to carry a packet of 11D information from nature. A Bloch sphere present in Fig. 4.4 may not show the entire brain but is a basic element, requiring additional balls for complex information (Fig. 4.5). The brain's mystery can be understood by sensors that return to the environment and use a prototype to ask an external agent, is it missing some components or can it fill the gap. Consider 47 regions of the Brodmann area in the cortex. Since the ratio of frequencies of cortical oscillators is not an integer, the nearby bands cannot phase-lock linearly. Even after it, oscillators of different bands combine with changing phases and give rise to a perpetual oscillation between transient stable phase synchrony and unstable. This is also true of the resonance chain [2]. The power between different layers in the brain suggests that perturbations at low frequencies that may cause power dissipation at high frequencies result in slow oscillations modulating fast neighboring events.

In many kinds of biomaterials, the resonance frequency changes or shifts. These shifts appear in the form of time. So we considered those times as time delays, oscillation time and connected those as a circle. We took that periodic behavior as a circle and clock assembly created by connecting those circles. Such transition or

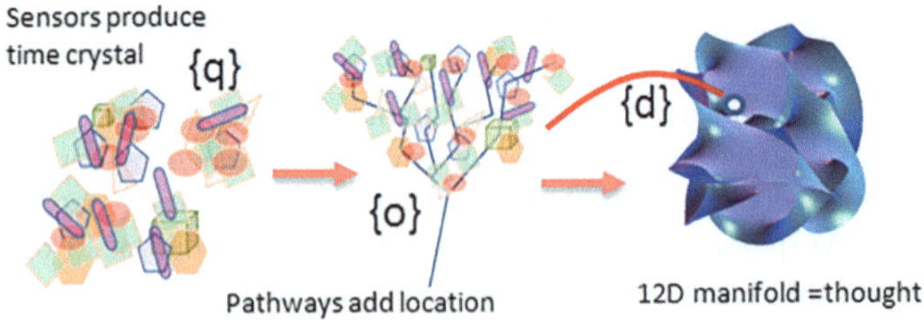

Fig. 4.4 This figure shows information propagation related to time crystals in three steps. A quaternion shows the time crystals of the sensors, the fusion time crystals through the pathways by octonion, the extension of information for twelve dimensions is shown by dodecanion [7]

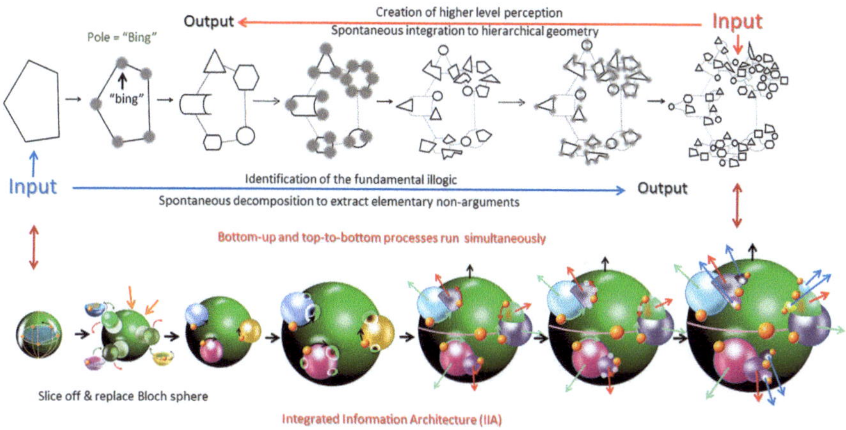

Fig. 4.5 The decomposition of a pentagon is shown in the top row. There are singularities at the corner points. Every point holds a geometric structure, while corresponding time crystals are shown in the bottom panel [7]

shifts appears in the resonance. We took 12 types of brain components. It is time taking process to build a whole structure from an elementary component. We built and simulated the 3D clock Assembly of all biomaterials. We measured the electric and magnetic field distribution and built 3D clock assembly models of 12 types of brain components. The 2D clock assembly model of the human brain has 537 clocks that are connected by phase and created a nested map.

4.4.3 A Map of 3D Clock Assemblies of a Cortical Column

In the 1970s, researchers considered that a single clock involves the time crystal. It was used to detect the life feature of neurons and viruses. Our brain controls all biological clocks of the body, which continuously reproduce the living cells. Cognitive features involve our brain run by biological features. We integrated such cognitive features in the forms of many tied running clocks (Fig. 4.6) in 2D form. The detected responses of the cortical column by us are an artifact or not. We studied the number of cognitive responses offered by the cortical column [66, 67]. All cognitive features of the brain are well documented in the literature [68], and we selected 20 well-known cognitive pathways. An operational machine-based clocking architecture of cortical column is shown in Fig. 4.6. We could build such clocks where such a set is the composition of AC signals that a resonant antenna could trigger. Such composition of clocks may offer many kinds of patterns which fellow GML, geometrical musical language [7, 10, 69] and PPM, phase prime matrices [11]. All clocks in the map are related to phase and could add a distinct type of dynamic features. The clocking map of a cortical column may alter. Some clocks may add or remove during the

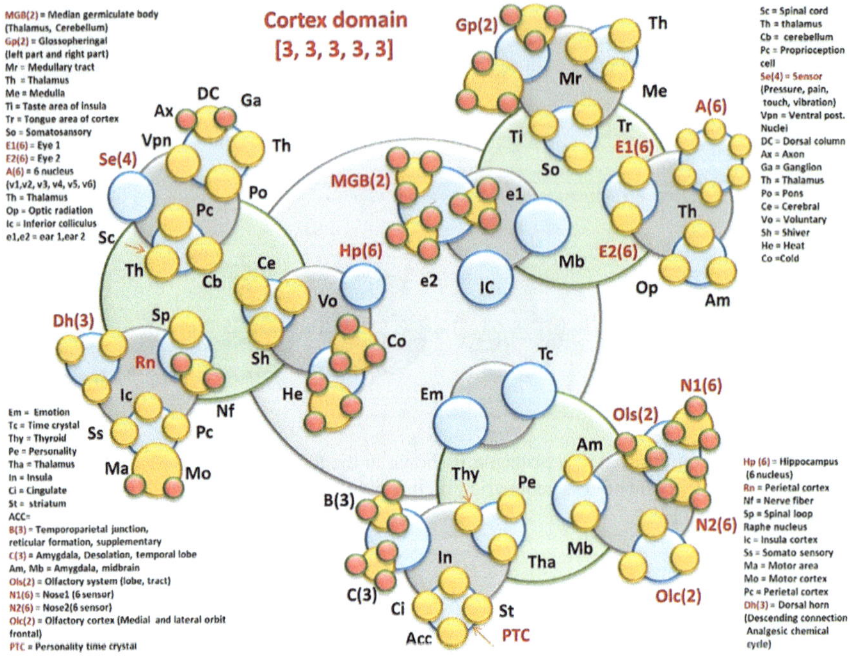

Fig. 4.6 2D clocking assembly model of cortex domain [7, 8]. We consider each circle as a clock. A guest clock on the circumference of a host clock means that the guest clock runs a loop. Geometric and self-assembled features have been associated with the cortical column. Following these features, we built the clocking assembly of the cortical column

decision-making process in the human brain. Clocking assembly could be a map to understand optical vortex patterns and EEG.

4.4.4 *Defects in Cortical Columns Assembly Are Fundamental to the Learning Process*

A cortical column collects many numbers neurons, and it is found perpendicular to the cortex layer. Neurons inside cortical column arranged in seven layers. Excitatory connections are found between neurons over a considerable distance. Due to these connections, the cortical column makes a rapid connection with the surrounding cortical column. Measurement of connection between them using non-invasive functional brain imaging is a little bit difficult. It shows the homogeneous properties in all directions. A particular spatial resolution range allows us to find out larger cortical columns.

Hexagonal symmetries of cortical columns are seen on the cortex layer of the brain and continuous till somatosensory cortex area. Barrel cortex of mouse studied in terms of symmetries of cortical column [70] while the functional significance of cortical column is still subject of discussion. Synaptic plasticity could be understood by a change in the lattice arrangement of the cortical columns [71]. Change in the hexagonal symmetry of the cortical column is key for memory storage and learning, which significantly changes the decision-making process in the human brain. In the 1993s, there was a huge argument that a single cortical column could function independently. Although, the collective responses of the cortical column would be the final output [72]. Neurons in the cortical column were found in asymmetrical order to transmit the signal with different phases. Here, we aim to determine the correlation between resonance frequencies and geometry of cortical column and find out the dependency of EM field on the neurons orientations.

4.4.5 Clocking Assembly Model of Major Components of Brain Components

A 3D assembly of clocks is a functional machine. If clocks run perpetually, it would continuously deliver output. Machine-like clocking models of major brain components like cortical domain, skin nerve net, cortical branches, microtubule, Neuron, thoracic nerve, thalamic body, cranial nerve, and blood vessel branches are depicted in Figs. 4.6, 4.7, 4.8, 4.9, 4.10, 4.11, 4.12, 4.13, and 4.14 respectively.

4.4.6 Blood Vessel Functional Responses: Motion or Movement, Audio + Visual + Time Registering an Event; Homeostatic Thermal Equilibrium

Our brain is the most complicated structure in the body, it is inside our head and is protected by the skull, and it is made by the billions of neurons or brain cells that are interconnected in such a way to allow us to do things like think, movement, smell, etc. (see Chap. 5). There are three major sections in the brain like the cerebrum, cerebellum, and brain stem. The brain stem is connected to the spinal cord. The cerebellum is divided into different lobes like the frontal lobe (thinking and consciousness), temporal lobe (sense of smell and sound), parietal lobe (pain, touch, and vibration), and occipital lobe (sense of size). All the brain lobes have to work together to perform the function listed in each lobe to specialize. That is brain anatomy. The brain is metabolically super active, so it needs lots of fuel like glucose, oxygen, and gets that from a rich blood supply. A recent study showed that the effective diameter of blood vessels changes from periodic vibrations of low frequency to lower extremities due to increased blood circulation by the vascular system.

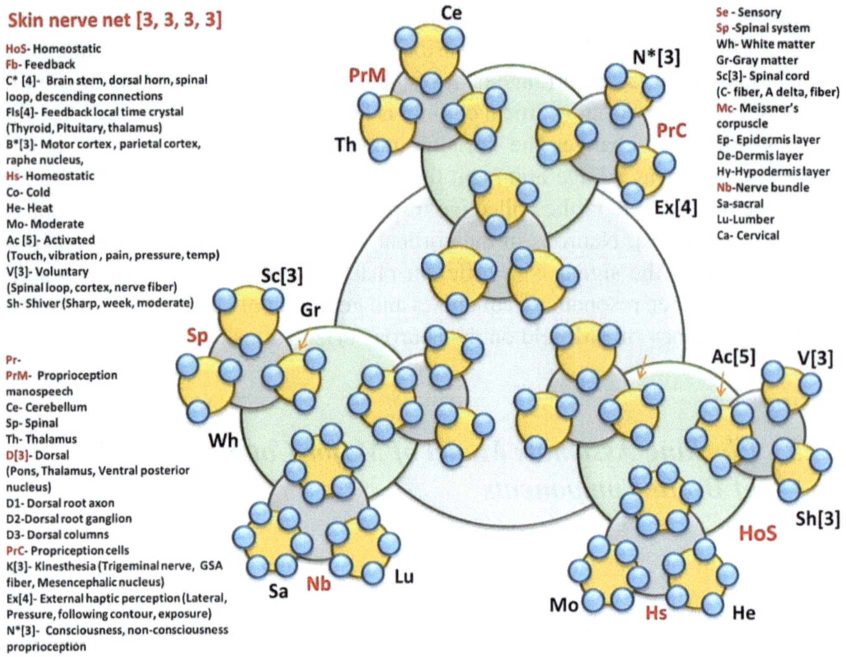

Fig. 4.7 2D clocking assembly model of skin nerve network [7, 8]

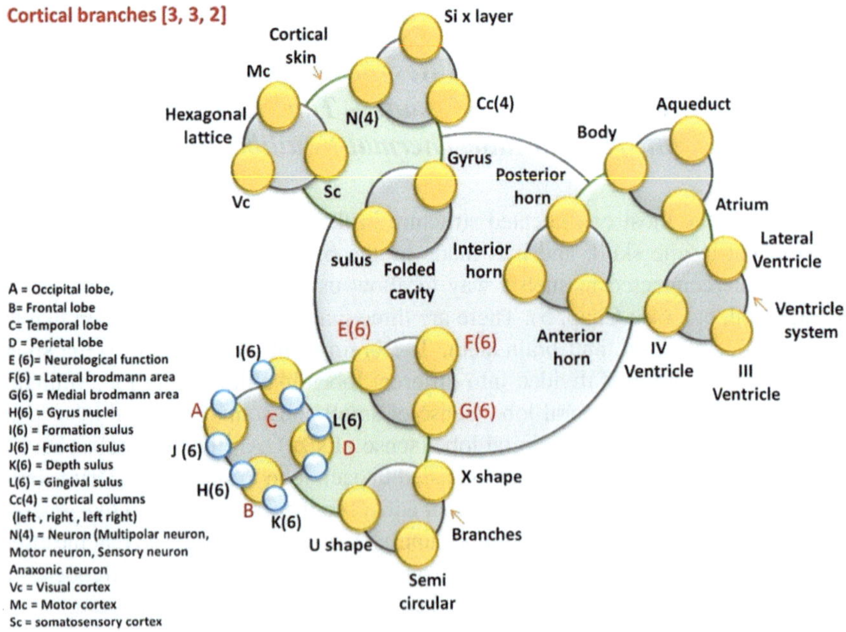

Fig. 4.8 2D clocking assembly model of human connectome [7, 8]

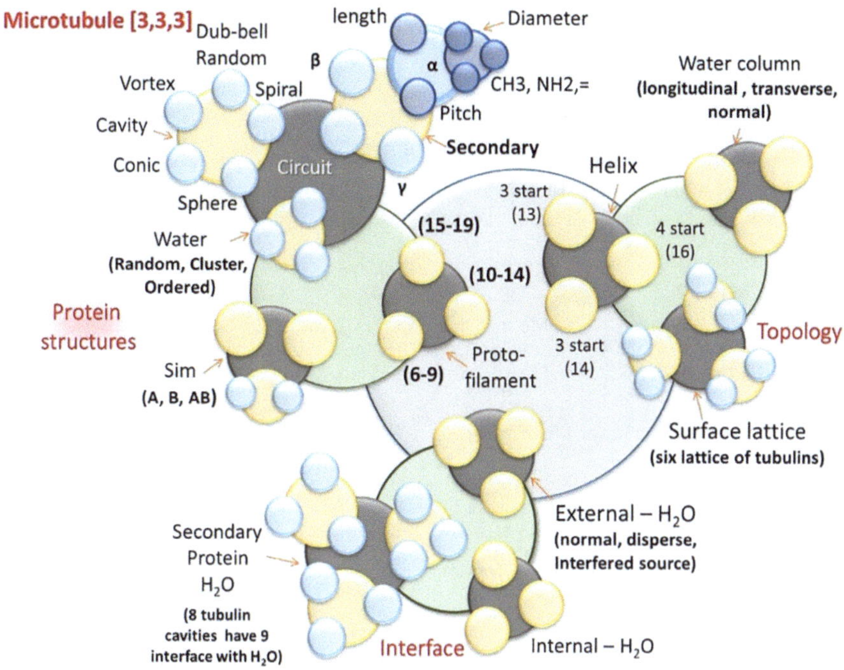

Fig. 4.9 2D clocking assembly model of microtubule [7, 8]

Blood flow rate and concentration and standard body temperature (37 °C) or homeostatic are controlled by the hypothalamic nucleus in the brain. Our skin has temperature-sensitive cells like sweat glands, skeletal muscles that sense temperature from the environment and send signals to the hypothalamus. The hypothalamus signal goes back to the skin, causing to increase or decrease sweating or shivering of the body when the temperature rises or falls from standard temperature. Over the past few decades, researchers have tried to find out how the vibrations of blood vessels affect the activities of our body. Mechanical vibrations in the body could be produced by footplate devices. Reported stimulation in the body increases the rate of blood flow. A mathematical model on the effect of body vibrations on blood flow in a single artery is described [73]. Researchers have tried to find out how the vibrations of blood vessels affect the activities of our body. Movement, homeostatic and thermal equilibrium are well related to the vibration of the blood vessel. Several studies show the effects of body vibrations on human activity. However, not a single study reports on the effects of mechanical vibrations of the blood vessel on consciousness features of the human body.

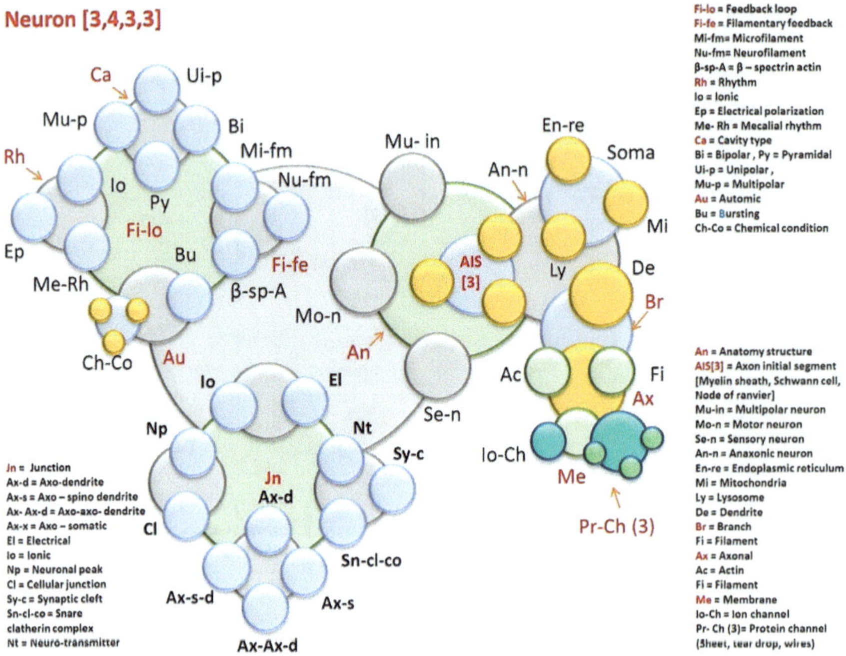

Fig. 4.10 2D clocking assembly model of neuron [7, 8]

Fig. 4.11 2D clocking assembly model of thoracic nerve [7, 8]

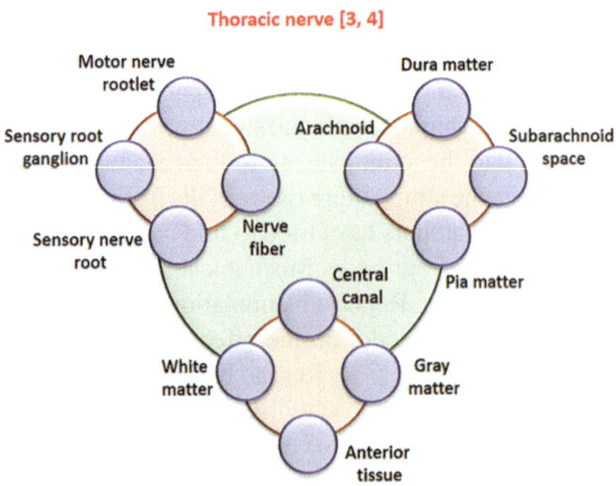

4.4.7 *Functional Map and Humanoid Bot*

The clocking map of mechanical rhythms of the blood vessel is shown in Fig. 4.14. In the vibrational state, rhythms of blood vessels triggered with its sub-components like thyroid, retina, basal ganglia, adrenal, melatomic level, hypothalamus, pituitary

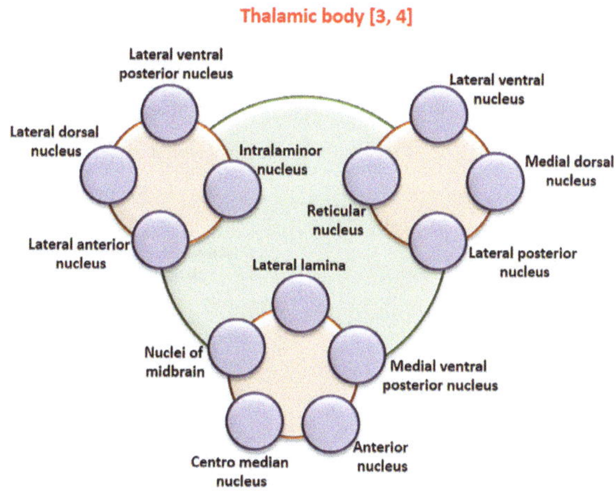

Fig. 4.12 2D clocking assembly model of thalamic body [7, 8]

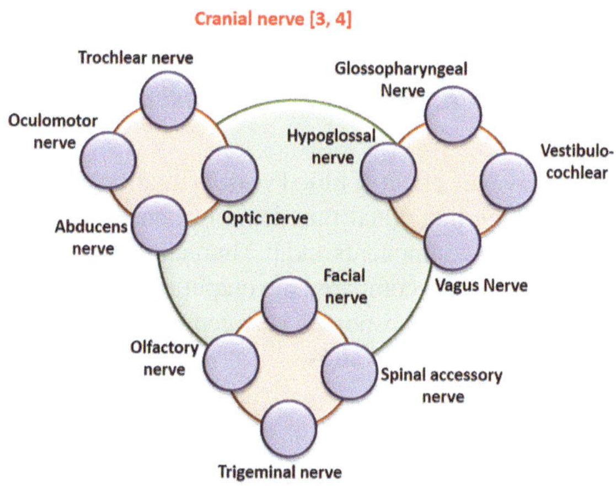

Fig. 4.13 2D clocking assembly model of cranial nerve [7, 8]

gland, gonad, thalamus with different periods, its neighbor's components perform their function at the resonance frequency. We represented a single rhythm by the single clock, and their size or diameter depends on the periods of vibration. The complication of the clocks regarding all constitutes generates the movement map. Similarly, the functional map of the Homeostasis and thermal equilibrium is the composition of known 30 types of clocks) interlocked by the phase and time and exhibited the function. In a practical case, we have built 20 types of well-known consciousness circuits in the brain of the humanoid bot (see Chap. 5), and the EEG machine confirms their outputs in the form of brainwave (alpha, beta, gamma, and delta) or neuron firing patterns.

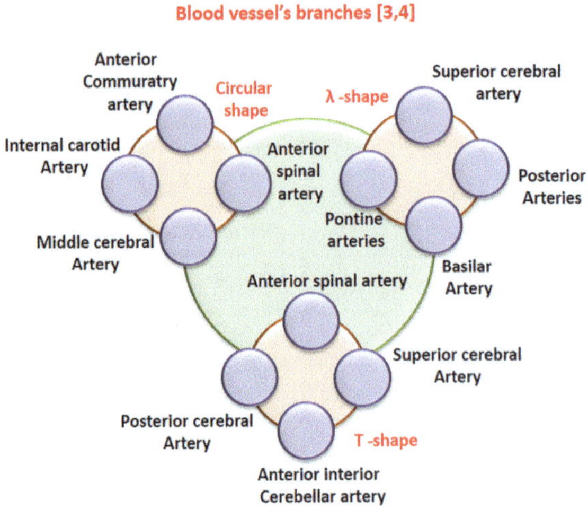

Fig. 4.14 2D clocking assembly model of blood vessel branches [7, 8]

4.4.8 *How Do Vibrations of Blood Vessels Affect the Experiences of Consciousness?*

To observe the effect of blood vessel vibration on connectome function, first, we have detected the individual functional responses of both organs. Finally, we integrated all the brain components and the humanoid bot' brain and observed the output in the brain waves. We compared the output of the connectome-based brain with the HBS brain. The output responses between individual structures correlate to each other in terms of frequencies ratios.

4.4.9 *Structural Symmetry*

Blood vessels and connectome are the structure of the nerve fibers and form the cavity shape. Arteries of the blood vessel supply the blood to the cortex region and appear in five symmetrical regions. The connectome is the projection of neuronal connection in the brain, which is identified in five symmetrical nerves bundles. These nerve bundles are distributed in five symmetrical regions on the cortex regions of C5 symmetry. In contrast, in each symmetrical region, nerves distribute in 3 symmetrical regions, or C3 symmetry is there. Connectome contains a triplet of pentate (C3 of C5) symmetry. Overall C5 symmetry in both. The internal configuration of structures may be important to influence the functions.

4.4.10 Overlapping Energy Pathways

During performing the various tasks, the human brain activates common components or pathways. For example, the thalamus is the central unit from where every pathway passes. For example, connectome and blood vessel offer the consciousness and movement feature, respectively. The circuital maps for both features contain the common components like thalamus, amygdala, and hippocampus in pathways. So small triggering in those components made an effective change in brain output.

4.4.11 Role of Homeostatic and Motion Circuits in HBS Brain

While making Humanoid bot subject (HBS), we have implemented 20 types of consciousness features in the HBS brain equivalent to the connectome output. HBS outputs are detected when HBS is triggered by external factors like the visual, touch, motion, sound, position, and temperature. The built homeostatic circuit maintains the temperature and the signal flow rate along with the nervous system of HBS, and finally, it refers to the signal in brain-sensing regions. Similarly, the motion or movement, audio + visual + time, registering an event of HBS are sensed by the built map in the HBS brain. The EEG machine maps corresponding outputs.

4.4.12 Detection of Consciousness Characteristics on the Brain of HBS by EEG

Twenty consciousness circuits are built in the HBS brain, and the Epoc EEG machine detects their output. To check the activation and deactivation circuits, we have triggered the HBS by offering one by one different external signals like visual (seeing some object), touch, motion (vibrating the HBS), sound (noise), position (change the position), and temperature (produced the heat near the HBS). The attached sensors detect all signals on the HBS body which are connected to the nervous system. Signals propagate from sensors to the nervous system to the HBS brain to the upper cortex layer mimicked by the PVC plastic. These circuits' activation and deactivation effects can be seen in the cortex layers in terms of brain waves/neuron firing patterns. Effective changes in the neuron firing patterns are seen by switching on and off of sensors. From we find that when all attached sensors are off, the neurons fire in an unusual way, a noisy profile appears. When the signal activates, arranged patterns of neurons appear in the respective regions of cortex layers. For example, if we put some colorful objects in front of the HBS eye, the random neuron firing pattern concentrates in the visual region. A similar thing happens with the touch, motion, sound, and position sensors. In the case of temperature, we have produced the heat near the temperature sensor. In that way, position and visual sensors both get activate resultant combined cortex regions blinks.

4.4.13 The Complication of All Possible Clocks in the Blood Vessel and Connectome Model in the Form of the Time Crystal

We have compiled all possible clocks used in replicating the vibrational activation performance of the blood vessel and connectome functions in the form of the time crystal. A time crystal is a well-defined concept made from the clocks and explains the life-like features. Sixteen clocks contain in blood vessel time crystal. Single clocks mean the periods of the signal that pass through that particular component. So the diameter of clocks depends on the periods. The initial signal passes through the circular, lambda, and T shape architecture of the blood vessel. There form three clocks, after that it in each of those regions. It further passes through four individual components (for example, in circular shape clocks, it passes through the Anterior commuratry artery, internal carotid artery, middle cerebral artery, and anterior spinal artery), so there are four clocks. Total number of clocks are $3 \times 4 = 12$ (Fig. 4.14). Connectome contains the 48 clocks during the performance of the connectome functions (Fig. 4.8). Such clocking architecture of the blood vessel and connectome may be useful to detect the future output by activation and deactivation of any clock or any patient with a blood vessel defect. From the BV function, we can easily detect which component or clock is not working.

4.4.14 How Do the Cerebellum, Hippocampus, and Hypothalamus Establish Instantaneous Communication During the Decision-Making Process?

There is a common communication mode between the cerebellum and hypothalamus. A signal goes to the hypothalamus through the cerebellum. Hypothalamus has 13 nuclei, and it set the resonance of those nuclei in a way that synchronous signals from the cerebellum and passes to the hippocampus. Hippocampus has the facility to synchronize the coming signal by circular loop channels and store them as memory.

It is possible to build hardware equivalent to brain components (cerebellum, hippocampus, and hypothalamus) using their clocking assembly model (Fig. 4.15a–c). If we built the circuits in the same way, it would follow the features of brain components. The clocking assemblies of major brain components are presented by tracing the clocks from brain standard books. The temporal map of the human brain

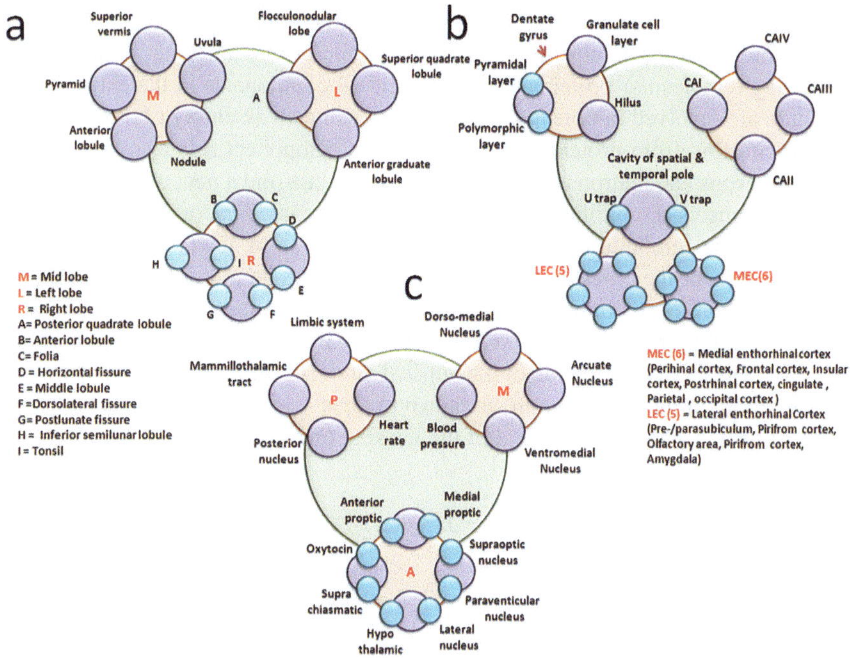

Fig. 4.15 Nested clocking maps of the cerebellum, hippocampus, and hypothalamus are shown in panels (**a**), (**b**), and (**c**), respectively

consists of 537 clocks. When the human brain makes the decision, all clocks run parallel.

We made efforts to build the communication link between the cerebellum, hippocampus, and hypothalamus at their resonance frequency. Replicated geometry is 100 times or 1000 times larger than the actual size and resonance in KHz or MHz frequency range. Built components in theoretical and experimental cases always maintain the ratio of the resonance frequency. The signal propagates between these components and finally transfer to the brain regions. The resonance frequency of the cerebellum is dual or half of hypothalamus resonance frequency, while hippocampus frequency is three times hypothalamus frequency. In summary, (1) brain components always maintain the ratio of resonance frequencies; however, their size does not matter. Their geometry is everything. (2) Resonance frequency of one component is an integral multiple of other brain components.

4.5 A Time Crystal Model of the Human Brain

The undefined nature of the human brain has prompted speculation that the brain acts as a black hole [25], creating a new concept to the time crystal map. For a long time, it has been an important point to the discussion of ‘how the human brain keeps time’ [74]. Keeping time is the basis of the brain’s information processing [75]. Single molecular proteins to neurons and all components in the entire neuron structure are involved in time management [76] using an assembly of clocks or time crystal. Starting from protein, every single brain component acts as a time crystal. Single neurons and neuron assembly like basal ganglia find a new trick; it forms the corresponding structure of the time crystal [77]. The same type of time management is found in the cerebellum when it detects coordination in movement; it makes the structure time or the clocks [78]. The action of the human motor keeps an inherent fractal structure of a similar pattern from seconds to minutes to hours. The fractal design is found throughout the brain in different kinds of literature. Still, there was no collection about the creation of the musical model of the brain. A 2D and 3D time crystal model of the human brain, as shown in Figs. 4.16 and 4.17, respectively will allow offering impeccable renewal of the missing organs [79].

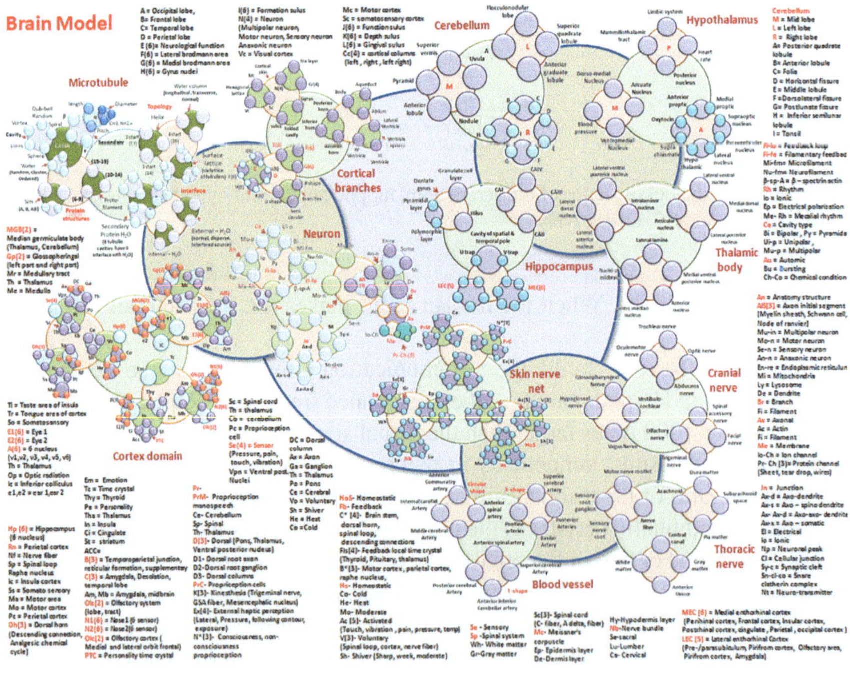

Fig. 4.16 Time crystal model of the human brain made of 537 classes of clocks [8]

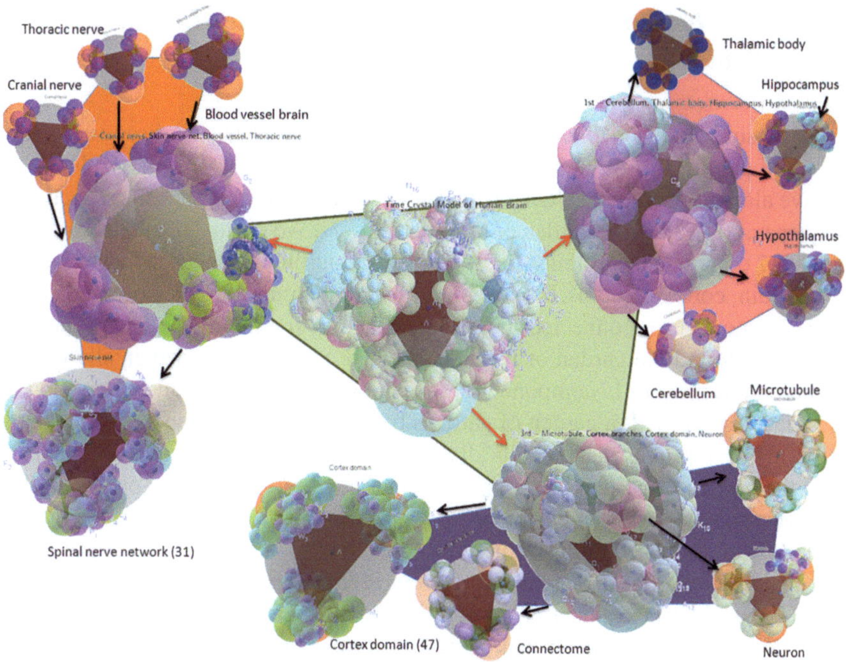

Fig. 4.17 3D spherical map of the time crystal of the human brain [8]

4.5.1 Twelve Nested Layers Within and Above Constitute the HBS Brain

Sincerely, we have built almost all the components of the human brain in electromagnetic simulation software and its hardware for the experiment. We have considered solid shape as dielectric and hollow shape as a cavity resonator. Maxwell equations solved all components in the time domain solver. The results obtained from the components of the brain have been verified theoretically and experimentally. A complete-time crystal map of the human brain is the complication of 537 classes of clocks. We have traced all the clocks from the database reported for the last 80 years. Five hundred thirty-seven classes of brain rhythms are reverse engineered by a single cable of the wire using a 1:1 million ratio and 20 conscious circuits (Consciousness; Memory; Language and conversation; Thinking and Intelligence; Sense of universal time, symmetry; Fear, threat, anger, hate; Reward; Mimicry, skill, adaptation; Creativity & Humor; Personality; Love and pain; Learning dreaming defragmentation; Fusion of elementary sensor into single time crystal; Proprioception; Motion or movement, audio + visual + time; Homeostasis, thermal equilibrium; How brain senses direction; Temporal synchrony of entire skin cover, feeling of body; Emotion; Synthesis of time) are replicated in the brain of humanoid (HBS-Human Bot Structure) that shows a human-like response (see Chap. 6).

4.5.2 An Operational Time Crystal Model of the Human Brain

An extensive study on the origin of biological rhythms led to discovering time crystals operating in the human brain and the whole body. Time crystals do not survive alone; a biological clock creates a set of many time crystals as a function of frequency. It is called the garden of meander flowers. We created the dielectric structure of 12 components proteins, microfilaments, microtubule, neuron, nerve bundles in the brain, cortex columns, hippocampus, hypothalamus, thoracic nerves, thalamic bodies, motor neuron connected skin structure, and by simulating the electromagnetic resonance, found the garden of meander flowers using our experimentally derived Hamiltonian for every 12 components. Since the angle between meander petal flowers is 2π/integer, we built a garden of gardens by combining 12 components as a brain and a metric of primes that predicts all possible time crystals that one could generate from the garden of garden that represents a human brain. Using 3D printed dielectric structures, we verified our theoretical prediction and developed an operational model of the human brain as a drive-through isochrone [38] around phase singularity created by primes. Generalization of integer patterns used by meander flowers of time crystal leads to a phase prime metric. It paves the way to design and build an integrated brain-like decision-making architecture using market-available clocks.

In the 1970s, the time crystal meant a 3D spiral of the geometric phase where three orthogonal axes are the input signal phase, the output signal phase, and the normalized perturbation converted into phase. Randomly applying a suitable perturbation to a clock does not create a time crystal; a critical condition must be met. A shift in that parameter creates a "landscape of meander flower or time crystals," where the dimensionless ratios of excitation to recovery time scales ε were plotted with order parameter k or frequency. Dielectric resonance of biomaterials are being studied since the 1930s, the geometry of a dielectric structure regulates the electromagnetic resonance band, and such studies were made in the 1950s. Thus, we generated theoretically the possible time crystals of all 12 systems. Using an enormous brain research database contributed by thousands of scientists in the last century, we have integrated the clocks into a singular operational architecture. Neuron, microtubule, and protein data are experimental time crystals generated by us in our laboratory. The time crystals for the rest of the systems are collected from various research papers. Now, we try to match dielectric resonance derived time crystal with the biological observations, matches, and discrepancies outlined here, which would be gradually resolved in the future. The meander flower gardens are constructed for all 12 systems, and we found how these architectures select the integers while designing the biological components. When we combine 12 systems that are physically one inside another, we get a singular garden of gardens wherein we find unique routes around dM. The distinctness of the 12 systems is separated by dC (Hopf bifurcations). We have proposed a generic Hamiltonian to generate the dynamics of each 12 systems and for the integrated system as a single unit.

4.6 Dodecanion Algebra is the Mathematics of the Brain and Our Universe

Geometrical algebra is used to link integers by geometrical shapes [80]. Algebra already exists for certain imaginary numbers, for example, quaternion algebra for three imaginary worlds and octonion algebra for seven imaginary worlds. All types of algebra were related to Clifford algebra and Lie algebra. Here, we have addressed the complex transition from octonion algebra to dodecanion algebra at a large scale for 11 imaginary worlds [31, 32]. One can construct tenors for dodecanian algebra (Fig. 4.18), a primitive tensor in multinomial algebra or variable numbers of imaginary worlds. Continuous fractal geometric algebra (CFGA) is used in mathematical calculations, a new type of mathematics; for example, if you draw a circle, an overall mathematical process will be used to convergence that drawn circle. For the artificial brain model, a new operator is suggested, a geometric operation of ten steps sequentially and solves all the necessary mathematical functions.

The fundamental idea of current brain-building projects suggests that if one mimics a structure like a neuron, its properties are copied by themselves, logically this is not true because the matter is isolated, discrete, and complete, but it is not a time crystal. If both (the structure of the matter and its properties) are not copied, then we lose information about the 'happening events' of the mimicked structures. Time crystal structure reflects its dynamics of the future. It is the seed that is transplanted

Dodecanion h_j

h_i

h_{ij}	h_0	h_1	h_2	h_3	h_4	h_5	h_6	h_7	h_8	h_9	h_{10}	h_{11}
h_0	h_0	h_1	h_2	h_3	h_4	h_5	h_6	h_7	h_8	h_9	h_{10}	h_{11}
h_1	h_1	$-h_0$	h_3	$-h_2$	h_5	$-h_4$	h_7	$-h_6$	h_9	$-h_8$	h_{11}	h_{10}
h_2	h_2	$-h_3$	$-h_0$	h_1	h_2	$-h_7$	h_4	h_5	h_6	h_{11}	$-h_8$	$-h_9$
h_3	h_3	h_2	$-h_1$	$-h_0$	$-h_7$	h_2	h_5	h_4	h_{11}	$-h_6$	$-h_9$	$-h_8$
h_4	h_4	$-h_5$	h_2	h_7	$-h_0$	h_9	h_2	h_{11}	h_4	$-h_5$	h_6	h_7
h_5	h_5	h_4	h_7	$-h_2$	$-h_9$	$-h_0$	h_{11}	$-h_2$	h_5	h_4	h_7	h_6
h_6	h_6	$-h_7$	h_4	h_5	$-h_2$	$-h_{11}$	$-h_0$	h_9	h_6	h_7	h_4	h_5
h_7	h_7	h_6	$-h_5$	h_4	$-h_{11}$	h_2	$-h_9$	$-h_0$	h_7	h_2	h_5	h_4
h_8	h_8	$-h_9$	$-h_6$	$-h_{11}$	h_4	h_5	h_2	h_7	$-h_0$	h_1	h_2	h_3
h_9	h_9	$-h_8$	$-h_{11}$	h_6	h_5	h_4	h_7	h_2	$-h_1$	$-h_0$	$-h_3$	h_2
h_{10}	h_{10}	$-h_{11}$	h_8	$-h_9$	h_6	h_7	h_4	h_5	$-h_2$	h_3	$-h_0$	h_1
h_{11}	h_{11}	$-h_{10}$	h_9	h_8	h_7	h_6	h_5	h_4	$-h_3$	h_2	$-h_1$	$-h_0$

Fig. 4.18 Photograph of dodecanion tensor [31, 32]

into a phase-based matrix to form a tree. Traveling through singularity points shows a path to new Science.

Microtubules are grouped inside neurons and give them strength. A triple of triple resonance bands was observed. Proteins and neural membranes that form the microtubule explain the triplet of triplet character [2]. A scale-free-fractal feature that shows the special pattern of PPM [81]. Researchers reported that when neurons fire, a group of spikes are formed, and it repeats itself. The conclusion is that spikes are organized as a group of groups that hold geometric information [82]. The neuron membrane is doing all that; as stated in 1907, it needs to be seen again [83]. The integrate and fire model of the neuron [84] is a topic of discussion. Membrane potential and diffusion of ions regulate the foundation of neuroscience and the brain science of today. Thousands of filled nano-filaments within the protein remain unknown. Even now, the neuron is like a hollow structure, in which the flow of ions creates the potential on the skin, and all things in it are silent.

4.6.1 A Conscious System: Limitation of the Human Brain Copy

The characteristic of the time crystal model of the brain is based on the state of consciousness in which a component must have at least twelve triplets of triplet resonance groups, so three different compositions of groups coexist and operate independently. Since the triangle is a separate entity, complete with sets of $2 \times 2 \times 3$, $3 \times 2 \times 2$, $2 \times 3 \times 2$, it has better control over the three complete identities of consciousness. It has been found that the major elements of the brain follow this condition, such as DNA, neurons, proteins, and cortical columns. The experiment already established it to have a specific arrangement of 3D clocks. Here we have compiled the widely distributed report into a single list. All parts share one consciousness in creating the brain of consciousness [85]. The entire brain behaves as a timekeeper, which manages the oscillations of the crystals present in the time crystals of the brain through their frequency fractal system by providing precise motion. Cognitive behavior results from a time tuning that internal and external rhythms undergo precise changes at all spatial scales, from a few atoms to the centimeter-scale range, and produce PPM.

Sensory input and neural collections or pathways follow the definition of geometry and contribute significantly to a time crystal of high order. According to the current model, nerves are wires, but in time crystals, pathways form clocks by connecting the symmetry in the lower time range. A 3D clock model of the human brain is shown in Fig. 4.17. No software is required to operate it. PPM is a dodecanian covered by a set of quaternion concepts to help examine and respond to nature.

4.6.2 Origin of Geometric Explanation of Events

Two correlated events in nature form the third event because the triangular correlation of events is primary [86]. Reichenbach Price suggested the structure of the three arrows [87]: past-future, asymmetric time arrow, and cause-effect. The feature of this suggestion is that it needs microphysical symmetry to connect these three. Price [87] drew the geometries between events. Here, asymmetric time flows in the shape of symmetry or the pattern form of primes. Human cognition has no relation to the past in explaining the present and predicting the future. All are elements of an architecture that is PPM.

4.6.3 Phase Prime Metric: PPM

Phase prime matrix links the symmetry of correlated events in the universe and presents the events in a way that follows the pattern of primes. PPM describes how an event in nature will occur. PPM does not use an algorithm or rule or code to examine how an event will unfold in the future and those events that we have failed to notice in the past. Sub-events are interconnected by topology, e.g., triangle or pyramidal, etc. [88]. If we write it linearly, then we lose its importance. There is a specific type of topology within every corner point of the topology and inside it. Thus the journey through the corner points showing the sub-events is endless. Events are integrated within or above, not side by side. For example, if we include a corner point in a triangle, we get a quadrilateral. Here PPM is a new scientific element that correlates random events. Instead of human memory and imagination, PPM understands the creativity of numbers that creates a new prime. We believe that nature considers synthesizing new symmetry for the self-assembly of primes. As a result, it governs the universe. If we perform this process in the brain jelly, we can feel PPM like feature. It predicts the future of events. Prime starts at 2 and 3; $2 \times 2 = 4$, and $2 \times 3 = 6$ is an empty gap between 4 and 6, in which the prime 5 should be called. Similarly, Brain jelly covers 220 million Prime for N^{12} [89].

86% of the entire integers can be generated using 15 primes in infinite space. Therefore, infinite potential events can be defined by shapes of 15 primes that can read unknown primes as the nearest composition of 15 primes. As current computers cannot operate the geometrical structures of events, we need a PC that uses a set of primes to analyze the entire compositions in real-time and add them to groups. An ordinary computer cannot show this kind of process because it often requires (i) trial and error (10^{23} per unit time), chemical collisions in the beaker often mimic it. (ii) Fifteen primes have been chosen out of 220 million primes at a time. No code can interact with 220 million nodes at a time. This can occur in a beaker through the diffusion and self-assembles the clocks. (iii) It is impossible to wire everywhere, multilevel imaginary or phase correlations, inside and above, etc. [31, 32]. This type of wiring is possible in supramolecular synthesis. Since the input activities of the

15 primes are fed to synthetic jelly, where one can write a set of primes and see a group of primes, growth of jelly, re-assemble, re-collection, etc. The grouped input of the primes connects events accurately. Organic jelly does not follow instructions but synthesizes outputs or decisions.

4.7 Fourth Circuit Element or Hinductor

Nye [90] suggested polarizing the monochromatic light and producing dark lines in the place where the electric or magnetic part of the light disappeared. If the electric field E is equal to 0, then in the 3D space, we find magnetic lines arranged like fibers. The flowing vortices of pure magnetic lines are not atoms. It is necessary to transform them into vortices. Only then we can use them as atoms. Interference between electric and magnetic fields is possible by the different symmetry of lattices. One can see free-particle-like structures made up of magnetic fields, some of them appearing as loops, curved, and some spiral. So this is the main transition in the study of [90]. Synthesis of pure magnetic lines is not possible in open space; perturbation is inserted through suitable devices. Lines that form on a triplet of cylinders then come out of the surface like vortex particles later used for different purposes. We can change the interference state of the spiral cylinder by charge and create a clockwise or anti-clockwise rotating free magnetic ring structure. In the phase space, the device with three concentric spiral cylinders appears like a hollow sphere consisting of twelve holes that continuously open and close. The hole indicates the particular length (time-span), diameter, pitch, and area of a lattice, from which we cannot get any defined output. The dark lines in phase space enter from one side while the exit from the other side. The projection of lines occurs by the phase gap. The phase gap connects the lines to the loops and creates a clocking path for the nested loops. The vortex forms a combined form of atoms by the superposition of multiple loops. A particular vortex ring may be an atom, but the combination of the loops is like a crystal. In the discussion, we mentioned that the charge flux device is identified as the fourth circuit element. Its outcome is not new. However, due to the second contradictory properties of the fourth circuit element, we have known this device as Hinductor (Fig. 4.19).

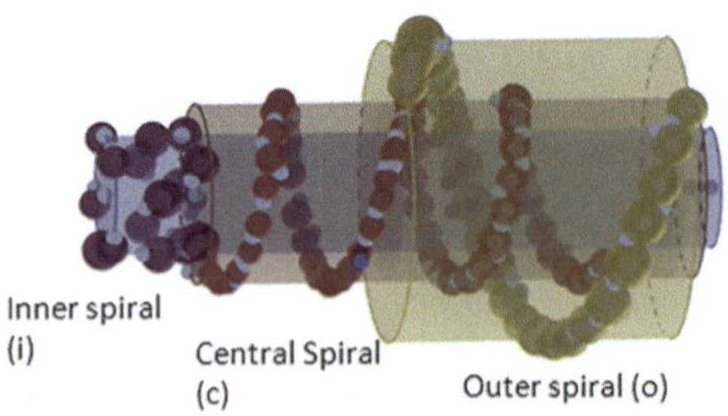

Fig. 4.19 Fundamental device (H) with three concentric spiral layers [89]

4.7.1 *Vortex Synthesizer: The Fundamental Engineering of the Brain*

When polarized monochromatic light is incident on the Hinductor, the stored pattern of the charges distribution on its surface can show different periodic oscillations. Monochromatic light incident to read what is printed on the surface of the device. The distributed charge forms pure magnetic lines in the dark. Understanding the structure of atoms created by magnetic light and soon identifying how effective the pattern of charges on a device is a story in itself. Monochromatic light does not destroy the oscillations of charge or coupled clocks in the device. Monochromatic light examines patterns of charge, and their approximate patterns are observed on a magnetic sheet. In the presence of suitable instruments, the concentric spiral cylinders are converted into a triplet, in which time crystals are written. The Yagi antennas send electromagnetic signals to vibrate the concentric cylinders. The brain jelly synthesizes a Hindictor using the given input information. Jelly forms an H by absorbing the time crystals as nested vibrations, which collect H to form a higher level of H. This process goes on continuously.

4.7.2 *A Hinductor Self-Assembles to Create Another Hinductor: Integrate Brain Hardware*

The helical shape of the water channels is the smallest decision-making element in the brain that is present in proteins such as microtubules, DNA, etc. The decision-making element cannot work by itself. Hinductor (concentric spiral cylinders) is a fundamental element of generating time crystals, as shown in Fig. 4.19. There are many such special structures in our brain-body. An example is shown in Fig. 4.20, microtubules. The microtubule's internal water channel, external water channel, and protein cylinder form a triad, showing a complete H device. Several H devices come close together to form a bundle in the form of an axon core. Actin-spectrin filaments wrap around that bundle and form a hollow cylinder. Then oligodendrocytes, astrocytes, and all branches form 3 layers. We find H2. Finally, many such H2 forms neural cortical columns. These millimeter-thick columns form the 2D film. The film holds different shapes (as H3 in Fig. 4.20) for the different elements over the brain-body system.

4.8 Conclusion: Global Database of Brain Rhythms

If polyatomic time crystal-based models of the human brain is a right perspective then we live with the past, present, and future at a time. Therefore, we are a creature of primes that hold symmetry of ordering of primes while creating integers, we see

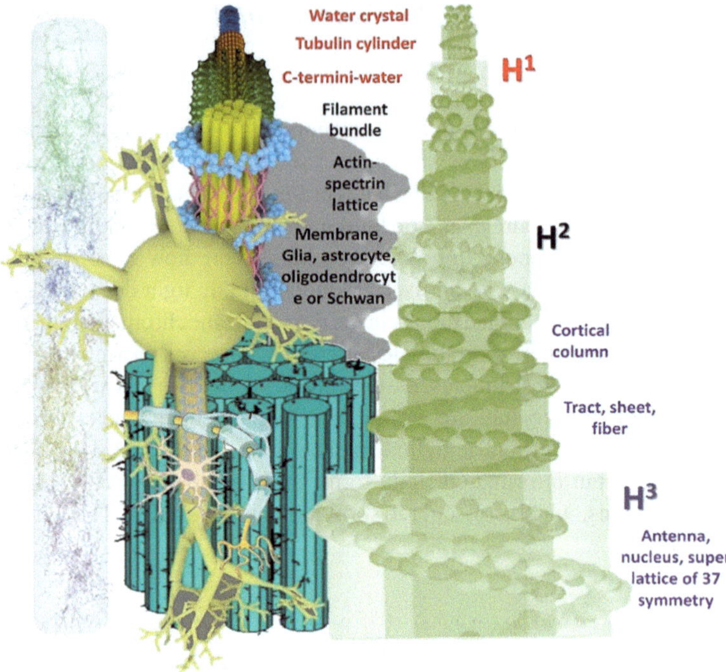

Fig. 4.20 The fundamental decision-making device in the brain is H, found in the vortex or spiral shapes such as alpha-helix, microtubule, etc. Two coated designs are presented. A coated structure of the basic element (H) such as cortical column found in the brain, is shown on the left side. A schematic diagram of the formation of the layered spiral is shown in the right side. The three films that make up triplet, are shown in red, black, and blue colors. H1 is the basic element that stores H2, and many H2 self-assemble to form H3 [89]

the changes as geometric transformation of world lines. We compiled past research studies and suggested a deep clock inside the neurons and a time map of a single protein. A complex mechanism has been described to find out the music of proteins and how a group of atoms lives over time as we live. Thus, we have created a global database of brain rhythms. The organic nano-machines are inspired by the complied proteins that realize the formation of a clock. Our perception is reflected in femtosecond clocks of some atoms of proteins to the nanosecond clocks of the protein complex to a millisecond clock of neurons in a hundred-year clock. Our database will be released soon and would help medical experts diagnosing by seeing the active and inactive clocks. The time-bandwidth of our conscious experience is in about 10^{26} orders. A brain is more than a time machine [25]. At all scales, all clocks simulate the past, present, and future. All clocks interact in real-time. Until we understand how nature assembles clocking architecture using a matrix that has no rule, no boundary, and no assumption to build, we cannot think about the time and how any material breaks symmetry to keep time. So we studied to learn about the physics and mathematics of time and learn how organic reactions can synthesize time as clocks. Elementary particles such as loops and knots of the energy transmission path follow the symmetries of the primes. To understand how the universe uses primes, we presented here a 3D clocking map of the human brain, the most fundamental prototype of the universe.

References

1. Ghosh S et al (2014) Design and construction of a brain-like computer: a new class of frequency-fractal computing using wireless communication in a supra-molecular organic. Inorg Syst Inf 5(1):28–100
2. Saxena K et al (2020) Fractal, scale free electromagnetic resonance of a single brain extracted microtubule nanowire, a single tubulin protein and a single neuron. Fractal Fract 4(2):11
3. Green AE et al (2017) Thinking cap plus thinking zap: tDCS of frontopolar cortex improves creative analogical reasoning and facilitates conscious augmentation of state creativity in verb generation. Cereb Cortex 27(4):2628–2639
4. Nie Y, Fellous JM, Tatsuno M (2014) Information-geometric measures estimate neural interactions during oscillatory brain states. J Front Neural Circ 24(8):11. https://doi.org/10.3389/fncir.2014.00011
5. Tank DW, Hopfield JJ (1987) Collective computation in neuron-like circuits. Sci Am 257:104–115
6. Rabinovich M, Huerta R, Laurent G (2008) Neuroscience. Transient dynamics for neural processing. Science 321:48
7. Bandyopadhyay A (2020a) Nanobrain: the making of an artificial brain from a time crystal. Taylor & Francis Inc. Imprint CRC Press Inc., Bosa Roca, United States, p. 36. ISBN 10-1439875499. ISBN 13-9781439875490. https://doi.org/10.1201/9780429107771
8. Singh P et al (2020) A self-operating time crystal model of the human brain: can we replace entire brain hardware with a 3D fractal architecture of clocks alone? Information 11(5):238
9. Gerisch G, Hulser D, Malchow D, Wick U (1975) Cell communication by periodic cyclic-AMP pulses. Phil Trans Roy Soc Lond B 272:181–192
10. Bandyopadhyay A, Ghosh S, Fujita D (2020b) Universal geometric-musical language for big data processing in an assembly of clocking resonators, JP-2017-150171, 8/2/2017: world patent, WO 2019/026983; US Patent App. 16/635,900
11. Bandyopadhyay A, Ghosh S, Fujita D (2020c) Human brain like intelligent decision-making machine; JP-2017-150173; 8/2/2017; World patent WO 2019/026984; US Patent App. 16/635,892
12. Bandyopadhyay A, Fujita D (2021a) Method for realizing quantum cloaking in electromagnetic device for remote imaging apparatus; Application no. 2021-172701
13. Bandyopadhyay A, Fujita D (2021b) Electromagnetic device, magnetic and electrical vortex synthesis device and magnetic and optical vortex synthesis device; Application no. 2021-172702
14. Shannon RV, Zeng F, Kamath GV, Wygonski J, Eke-lid M (1995) Speech recognition with primarily temporal cues. Science 270:303–304
15. Brown S, Martinez MJ, Parsons LM (2006) Music and language side by side in the brain: a PET study of the generation of melodies and sentences. Eur J Neurosci 23(10):2791–2803. https://doi.org/10.1111/j.1460-9568.2006.04785.x
16. Schwartz DA, Howe CQ, Purves D (2003) The statistical structure of human speech sounds predicts musical universals. J Neurosci 23:7160–7168
17. Koppelmans K (2016) Brain structure plasticity with spaceflight. npj Microgr 2:2
18. Rusov VD et al (2012) Can Resonant oscillations of the earth-ionosphere influence the human brain biorhythm? Gener Phys 1–13

19. Rensing L, van der Heiden U, Mackey MC (eds) (1987) Temporal disorder in human oscillatory system. In: Springer series in synergetics, vol 36. Springer, New York
20. Berliner MD, Neurath PW (1965) The rhythms of three clock mutants of ascobolusimmersus. Mycologia 57:809–817
21. Andrew C, Pfurtscheller G (1996) Event-related coherence as a tool for studying dynamic interaction of brain regions. Electroencephalogr Clin Neurophysiol 98:144–148
22. Reynolds JH, Desimone R (1999) The role of neural mechanisms of attention in solving the binding problem. Neuron 24:19–29
23. Aschoff J (1965) Circadian rhythms in man. Science 148:1427–1432
24. Aschoff J, Wever R (1976) Human circadian rhythms: a multi oscillatory system. Fed Proe 35:2326–2332
25. Buonomano D (2017) Your brain is a time machine. In: The neuroscience and physics of time. W. W. Norton & Company Inc.
26. Buzsáki G (2006) The rythms of the brain. Oxford University Press
27. Buzsáki G, Bragin A, Chrobak JJ, Nádasdy Z, Sík A, Ylinen A (1994) Oscillatory and intermittent synchrony in the hippocampus: Relevance for memory trace formation. In: Buzsàki G, Llinás RR, Singer W, Berthoz A, Christen Y (eds) Temporal coding in the brain. Springer, Berlin, pp 145–172
28. Bub G et al (1998) Bursting calcium rotors in cultured cardiac myocyte monolayers. PNAS 95:10283
29. Jirsa V, Haken H (1996) Field theory of electromagnetic brain activity. Phys Rev Lett 77:960–963
30. Ramkumar P, Parkkonen L, Hari R, Hyvärinen A (2012) Characterization of neuromagnetic brain rhythms over time scales of minutes using spatial independent component analysis. Hum Brain Mapp 33(7):1648–1662. https://doi.org/10.1002/hbm.21303 Epub 2011 Sep 13
31. Singh P et al (2021a) A space-time-topology-prime, stTS metric for a self-operating mathematical universe uses dodecanion geometric algebra of 2–20 D complex vectors. In: Ray K, Roy KC, Toshniwal SK, Sharma H, Bandyopadhyay A (eds) Proceedings of international conference on data science and applications. Lecture notes in networks and systems, vol 148. Springer, Singapore
32. Singh P et al (2021b) Quaternion, octonion to dodecanion manifold: stereographic projections from infinity lead to a self-operating mathematical Universe. In: Singh P, Gupta RK, Ray K, Bandyopadhyay A (eds) Proceedings of international conference on trends in computational and cognitive engineering. advances in intelligent systems and computing, vol 1169. Springer, Singapore
33. Buzsáki G (1989) Two-stage model of memory trace formation: a role for "noisy" brain states. Neuroscience 31:551–570
34. Bullmore E, Sporns O (2009) Complex brain networks: graph theoretical analysis of structural and functional systems. Nat Rev Neurosci 10(3):186–198
35. Chiel HJ, Beer RD (1997) The brain has a body: adaptive behavior emerges from interactions of nervous system, body and environment. Trend Neurosci. 20:553–557
36. Llinás RR, Ribary U, Joliot M, Wang XJ (1994) Content and context in temporal thalamocortical binding. In: Buzsáki G, Llinás R, Singer W, Berthoz A, Christen Y (eds) Temporal coding in the brain. Springer, Berlin, pp 251–272
37. Bennett BM, Hoffman DD, Prakash C (1989) observer mechanics: a formal theory of perception. Academic, San Diego, California
38. Pinsker HM (1977) Aplysiabursting neurons as endogenous oscillators. I: phase response curves for pulsed inhibitory synaptic input. and II. Synchronization and entrainment by pulsed inhibitory synaptic input. J Neurophys 40:527–556
39. Ringo JL (1991) Neuronal interconnection as a function of brain size. Brain BehavEvol 38:1–6
40. Richmond BJ, Optican LM, Gawne TJ (1989) Neurons use multiple messages encoded in temporally modulated spike trains to represent pictures. In: Kulikowski JJ, Dickenson CM (eds) Seeing contour and colour. Pergamon Press, New York, pp 705–713

41. Pierce JR (1961) Symbols, signals and noise: the nature and process of communication. Harper and Brothers, New York
42. Reddy S et al (2018) A brain-like computer made of time crystal: could a metric of prime alone replace a user and alleviate programming forever? In: Studies in computational intelligence, vol 761. Springer Nature Singapore Pvt. Ltd, pp 1–44. ISBN: 978-981-10-8048-7. https://doi.org/10.1007/978-981-10-8049-4_1
43. Yates FE (ed) (1987) Self-organizing systems: the emergence of order. Plenum, New York
44. MacKay RS (1991) Transition of the phase-resetting map for kicked oscillators. Physica D 52:254–266
45. Malinowski JR, Laval-Martin DL, Edmunds LN Jr (1985) Circadian oscillators, cell cycles, and singularities: light perturbations of the free-running rhythm of cell division in Euglena. J Comp Physiol 155B:257–276
46. Matthews PC, Strogatz SH (1990) Phase diagram for the collective behavior of limit-cycle oscillators. Phys Rev Lett 64:1701–1704
47. Wang XJ (1994) Multiple dynamical modes of thalamic relay neurons: rhythmic bursting and intermittent phase-locking. Neuroscience 59:21–31
48. Marder E, Calabrese RL (1996) Principles of rhythmic motor pattern generation. Physiol Rev 76:687–717
49. Mehta MR, Lee AK, Wilson MA (2002) Role of experience and oscillations in transforming a rate code into a temporal code. Nature 417:741–746
50. Ratte' S et al (2015) Subthreshold membrane currents confer distinct tuning properties that enable neurons to encode the integral or derivative of their input. Front Cell Neurosci 8:452
51. Radman T, Su Y, An JH, Parra LC, Bikson M (2007) Spike timing amplifies the effect of electric fields on neurons: implications for endogenous field effects. J Neurosci 27:3030–3036
52. Yamins DLK, DiCarlo JJ (2016) Using goal-driven deep learning models to understand sensory cortex. Nat Neurosci 19:356–365
53. Penttonen M, Buzsáki G (2003) Natural logarithmic relationship between brain oscillators. Thalamus Relat Syst 48:1–8
54. Buzsáki G, Chrobak JJ (1995) Temporal structure in spatially organized neuronal ensembles: a role for interneuronal networks. CurrOpinNeurobiol. 5:504–510
55. Eagles TH (1885) Constructive geometry of plane curves. Macmillan, pp 348–354
56. Collins B (2000) Visualization: from biology to culture. In: Bridges proceedings, pp 309–315
57. Halpern AR (2006) Cerebral substrates of musical imagery. Ann N Y Acad Sci 930:179–192. https://doi.org/10.1111/j.1749
58. Steriade M, Gloor P, Llinás R, Lopes da Silva FH, Mesulam MM (1990) Basic mechanisms of cerebral rhythmic activity. Electroencephalogr Clin Neurophysiol 76:481–508
59. Schmidt LA, Trainor LJ (2001) Frontal brain electrical activity (EEG) distinguishes valence and intensity of musical emotions. Cogn Emot 15(4):487. https://doi.org/10.1080/02699930126048
60. Sakai K, Hikosaka O, Nakamura H (2004) Emergence of rhythm during motor learning. Trends Cogn Sci 8:547–553
61. Glass L (2001) Synchronization and rhythmic processes in physiology. Nature 410:277–284
62. Glass L, Mackey MC (1988) From clocks to chaos: the rhythms of life. Princeton University Press, Princeton, NJ
63. Goila A, Pawar M (2009) The diagnosis of brain death. Indian J Crit Care Med 13(1):7–11 (The blue brain project). https://bluebrain.epfl.ch/
64. Kucewicz MT et al (2014) High frequency oscillations are associated with cognitive processing in human recognition memory. https://doi.org/10.1093/brain/awu149
65. Worrell et al (2008) High-frequency oscillations in human temporal lobe: simultaneous microwire and clinical macroelectrode recordings. Brain 131(Pt 4):928–937. https://doi.org/10.1093/brain/awn006. Epub 2008 Feb 7
66. Carter R (2014) The human brain book: an illustrated guide to its structure, function, and disorders. DK; United Kingdom, London
67. Zeki S, Shipp S (1988) The functional logic of cortical connections. Nature 335:311–317

68. Achard S, Salvador R, Whitcher B, Suckling J, Bullmore E (2006) A resilient, low frequency, small-world human brain functional network with highly connected association cortical hubs. J Neurosci 26:63–72
69. Agrawal L et al (2016) Fractal information theory (FIT) derived geometric musical language (GML) for brain inspired hypercomputing. In: Kacprzyk J (ed) Advances in intelligent systems and soft computing AISC, Series (Proceedings of SocTA), pp 37–61. Springer. ISSN: 2194-5357
70. Woolsey and van der Loos (1970) The structural organization of layer IV in the somatosensory region (SI) of mouse cerebral cortex. The description of a cortical field composed of discrete cytoarchitectonic units. Brain Res 17(2):205–242. https://doi.org/10.1016/0006-8993(70)90079-x
71. Horton JC, Adams DL (2005) The cortical column: a structure without a function. Philos Trans R Soc Lond B Biol Sci 360(1456):837–862
72. Gawne TJ, Richmond BJ (1993) How independent are the messages carried by adjacent inferior temporal cortical neurons? J Neurosci 13:2758–2771
73. Mishra JC, Pal B (1999) A mathematical model for the study of the pulsatile flow of blood under an externally imposed body acceleration. Math Comput Model 29(1999):89–106
74. Buhusi CV, Meck WH (2005) What makes us thick? Functional and neural mechanisms of interval timing. Nat Rev 6. http://www.nature.com/nrn/journal/v6/n10/abs/nrn1764.html
75. Wing AM (2002) Voluntary timing and brain function: an information processing approach. Brain Cogn 48:7–30
76. Ivry RB, Spencer RM (2004) The neural representation of time. Curr Opin Neurobiol 14:225–232
77. Graybiel AM (2005) The basal ganglia: learning new tricks and loving it. Curr Opin Neurobiol 15:638–644
78. Thach WT (1998) A role for the cerebellum in learning movement coordination. Neurobiol Learn Mem 70:177–188
79. Biasiucci A et al (2018) Brain-actuated functional electrical stimulation elicits lasting arm motor recovery after stroke. Nat Commun. https://doi.org/10.1038/s41467-018-04673-z
80. Hestenes D (1986) A unified language for mathematics and physics. In: Chisholm JSR, Commons AK (eds) Clifford algebras and their applications in mathematical physics, NATO ASI Series (Series C), vol 183. Springer, pp 1–23
81. Adhikari SK, Tomio L (1982) Efimov effect in the three-nucleon system. Phys Rev C 26(83) (Afimoveffect). http://www.wired.com/2014/05/physicists-rule-of-threes-efimov-trimers/?mbid=social_twitter
82. Curto C (2017) What can the topology tell us about the neural code? Bull Am Math Soc 54(1):63–78. https://www.quantamagazine.org/mathematician-carina-curto-thinks-like-a-physicist-to-solve-neuroscience-problems-20180619/
83. Brunel N, Van Rossum MC (2007) Lapicque's 1907 paper: from frogs to integrate-and-fire. Biol Cybern 97(5–6):337–339
84. Lapicque L (1907) Recherchesquantitatives sur l'excitationélectrique des nerfs traitéecomme-unepolarisation. J Physiol Pathol Gen 9:620–635
85. Nowakowski TJ (2018) Building blocks of the human brain. Science 362(6411):169
86. Reichenbach H (1956) The direction of time. University of California Press, Berkeley
87. Price H (1996) Time's arrow and archimedes' point. Oxford University Press, Oxford (especially chs. 1–2 & 8–9)
88. Courtland R (2018) Bias detectives: the researchers striving to make algorithms fair. Nature 558:357–360
89. Bandyopadhyay A (2020) Chapter 2 FIT, and GML; Chapter 3 PPM; Chapter 7. A complete, integrated time crystal model of a human brain. In: Nanobrain. The making of an artificial brain from a time crystal. Taylor & Francis Inc. Imprint CRC Press Inc., Publication City/Country Bosa Roca, United States, p 372
90. Nye JF (1984) Polarization effects in the diffraction of electromagnetic waves: the role of disclinations. Proc R Soc Lond A 387:105–132

Chapter 5
The Making of a Humanoid Bot Using Electromagnetic Antenna and Sensors

5.1 Introduction: Brain as an Electromagnetic Engine

5.1.1 Microwave and Radiowave are Transmitting Through Our Brain Making Decisions?

Information regarding wireless communication enhances the number of constitutes devices like base stations and other devices. High frequency (millimeter wave, 6–100 GHz) leaves a health impact. The characterization of the biological material varies in terms of power density in the duration of frequency exposure. Although, the available study does not provide significant information for safety evaluation [1]. Currently, many devices are used for communication purposes which is part of our daily life. Such devices allow the data to transfer wirelessly through the electromagnetic radiation in the radio frequency range without using a cable. Wireless devices emit harmful radiations that affect the human body and are also responsible for Alzheimer's disease, Parkinson's disease, fetus effects, and enhancing the risk of cancer cells. The utilization of wireless devices increases day by day across over all world. Wireless devices like cell phones, headphones, Wi-Fi routers, Bluetooth devices, and mobile towers lead to many types of diseases like brain tumors, heart disease, eye problems, and immune systems. Distinct types of radiation with different frequencies (3 kHz–300 GHz) are used to connect the wireless device [2]. In recent years, the effects of EMF through mobiles and telephones on the human nervous system emerges as a new research domain. GSM (Global system for mobile telecommunication) communication categorized the signal effects on sleep and cardiovascular regulation. Human nervous system is most energetic part of our body [3]. Many

The original version of this chapter was revised. Reference 29 has been corrected. The correction to this chapter is available at https://doi.org/10.1007/978-981-16-9677-0_7

P. Singh et al., *Biological Antenna to the Humanoid Bot*, Studies in Rhythm Engineering,
https://doi.org/10.1007/978-981-16-9677-0_5

peoples are facing harmful effects of radiation on the heart, red blood cells, etc. [4]. We should avoid utilizing wireless devices [2, 5].

From the 1960s, studies regarding the effects of EMFs on the biological system have been reported from the world. Circulatory system disorders with neoplastic diseases are known as a major health problem in the world. It could arise and enhance by exposure to EM fields. Circulatory system disorders are a major cause of electric fields generated by the external magnetic field. The electromagnetic field may perturb the human nervous system and the function of the circulatory system. EMFs seem effective at the left ventricle and decrease down to cardiac rhythms. Again some studies illustrate the disorder in blood pressure regulation and cardiac rhythms. Electromagnetic effects on the circulatory system are not reported there. So, available studies are insufficient to track the circulatory system functions [6]. It is not surprising to enhance health effects by diffusion of the electromagnetic field in communal and occupational environments.

The technological benefit of humans leads to the development of communication and electrical technologies, increasing the use of the electromagnetic field. Radio frequency electromagnetic field (RF-EMF) affects central nervous system (CNS) nerve cells. Functional change in ion channel and nerve myelin or RF-EMF acts as a source of stress in a biological organism. The possible effects of RF-EMF have not been detected yet. Available data are not sufficient to explain the probable health risk. In 1966, there was a huge debate on how a biological system is affected by a low-level radio frequency electromagnetic field. Such debate concerns the health effects of EMF exposure. EMF exposure may lead to a negative effect. Researchers argue that 5G EMF exposure causes more negative impacts. Many articles explain EMF effects that probably enhance the risk of neurological disorders, genetic damage, cancer, and memory, and learning process. EMF harms not only humans but also the surrounding environment [7]. It is necessary to establish new limits that consider new exposure features derived from the biological effects of electromagnetic radiation instead of the energy-based specific absorption rate.

Generally, non-ionic radiation is considered a harmless signal source. An electromagnetic pulse is a short burst of electromagnetic energy. 5G signal utilizes the higher frequencies signal with high magnitude/second. In the case of 5G, the issue is the pulse rate. Pulsed EMF is more biologically active and harmful compared to non-pulsed EMF. Every wireless device communicates by an electromagnetic field. A smarter device means high emission of pulsed EMF. 5G wireless signal may have a power deficiency but continuously emit pulsed radiation. It can damage DNA cells causing cancer. Emitted radiation interrupts the blood-spinal cord by destroying the vascular endothelial cells that guide a composite injury reaction. EMF from wireless devices is the cause of enhancing oxidative stress, hyperglycemia, and impaired insulin secretion in rat pancreatic islets. Diabetes is trigged in rats at 2.4 GHz frequency signal exposure. The relationship between RF-EMF coverage and schwannoma of the heart in a mouse is analyzed [8]. The link between oxidative stress and RF-EMF exposure may be a further research topic.

5.1.2 *Electromagnetic Resonance Model of the Brain*

Many researchers believe that the human brain acts like circuits, and complex circuits could generate a human brain, and everything happens at the level of the neuron. Creating an artificial brain would be easier if somehow we could have mapped the human brain accurately. A computer cannot have life; the circuit is encoded from neurons to neurons. We are adding an artificial engine (phase prime matrix-PPM) to find the life-like features. In an artificial brain, fake perception [9] and false memory [10] are real. Here we have considered the artificial brain similar to the universe, which shows the similarity at symmetry points. Some researchers consider the brain as an oscillator based on critical points [11], while it is more like a composition of flute and tuning forks [12].

5.2 Different Brain Models and How Do They Differ from the Time Crystal Model

5.2.1 *Boltzmann Brain Model*

The Boltzmann's Brain is the oldest model of the self-conscious unit. It has been suggested that a group of atoms in the thermodynamic equilibrium may take a structure similar to the human Brain [13], so the past must be a reflection of the future [14]. The time taken by a system of microstates in some regions of phase space is proportional to the volume of those regions, so all microstates or decisions are variable over time. In an artificial brain model, all components are related to the singularity [12, 15, 16]. Maxwell equations connect them to ensure the path of singularity [17, 18].

5.2.2 *Selfish Biocosm and Anthropic Brain Model*

According to "Copenhagen's interpretation of quantum mechanics," an event does not occur if no one sees it. By observing an event, a known observer turns to the universe's conscious life (participatory anthropic principle; [19]), the universe is biocentric. Regular observation changes the basic constants that lead to life [20]. Time travel can determine the final cause by including the probabilities of all paths from the beginning to the present through the ten dimensions [21, 22]. Here, we considered a new way of a life-friendly universe. The phase prime metric can extrapolate events as clocks or time crystals and represent those clocks or time crystals as the ratio of primes, but when geometrical musical language or GML is included [23], the geometrical constraints present in the metric control the formation of fundamental constants. Phase prime matrix provides a non-computable geometric output, which is consciousness-centered [12, 16, 24].

5.2.3 Brain Model Based on Turing Concept

Based on the Turing concept, the brain model believes that it is possible to create a conscious brain by asking a question whose answer is 0 or 1. The entire universe is a series of one-dimensional events based on this concept. A bit has two phases, but the structure of time crystal is about multi-phases; phase, clocking, speed, geometry, the direction of rotation, channel pattern, imaginary layer interactions, frequency ratio, etc. The unit of information is as complex as a life in itself [15, 25]. The structure of time follows a pattern of symmetries, and it is a nonphysical structure confined in a layer. PPM is a temple-like architecture, which connects patterns of symmetry and it senses the existing universe. All phase' prime metrics correct the lack of symmetry created by conflicts and thus advance life, so the universe [12] tries to attain the highest computation every moment in the brain [26].

5.2.4 Bayesian Brain Antimony

Decision-making or consciousness is caused by the conditional probability of various events [27, 28]. All models in this category are based on Wheeler's concept.

Connectome: computer-based brain architecture: There are many supporters of the idea of uploading an accurate map of neurons and mapping. Integrated information theory fits into this category, considering the observation of probability [12].

5.2.5 Brain Model Based on Free Energy

The proposal of such a brain attends our attention to the high amplitude analysis of the brain. The energy profile suggests that the brain does not have a 3D (three dimensions) structure; rather, it is itself a structure of 4D (four dimensions) [29]. Consciousness in the Brain comes from the study of information in 4D.

5.2.6 A Matryoshka Brain Model

Based on the Dyson region, Robert Bradbury proposed an imaginary megastructure of the brain known as the Matriosca brain. This brain has a high computational capacity. It occurs in a fractal form [30] based on the composition of the primes from which the integers are made, not their physical view. A group of primes is hidden in the infinite range; for example, some mathematician groups of $2 \times 3 \times 3 \times 3 \times 5 \times 7 \times 131$ and $2 \times 3 \times 3 \times 3 \times 5 \times 7 \times 97$ do not get the attention of any mathematician, whereas the two groups of primes are the same [16]. As such, many of these major compositions occur when we try to find the divisors of integers. Reddy et al. [15] designed them and saw the pattern of the prime hidden in them.

5.2.7 Brain Model Based on Electromagnetic Resonance

In such a brain, the synchronization of the spike is generated by resonance. It completes all the coherent logic that comes from human observation [31].

Harmonic and dipole brain model: Another approach is to analyze the brain on a harmonic basis. By flattering the components of the brain [32], several studies have reported that the lateral ventricle can be explained in neuro-expansion by spherical harmonics. Who resonates in a cortical column, and how does it happen? To answer this, we operated 12 types of rhythms in the brain that resonate in 12 types of temporal bands in 12 types of the carrier with resonant frequencies are found in the triplet of triple form in each band (Table 5.1; [12]).

5.2.8 Relativistic Brain Model

C2 symmetry or bipolar oscillations are related to resonance at the brain's left, right, bottom, and top levels. Results of DNA molecules or alpha-helix show modes of consciousness [33]. Poznanski [34] suggested how his research reproduces the central set of hypotheses in the dipole neurology theory [35]. The C2 symmetry alone covers 50% of the decision. Here we suggested how to use only a finite number of primes to increase it to 99.99%.

5.2.9 Schrodinger or Quantum Brain Model

The neurophysiologic response [36] in the human brain requires the receptor to be organized. We need a reference wave-generating source, where one can recover information by a holographic method. In quantum and holographic brain models, information is grouped into finite points where the endless choices collapse. The model changes to different patterns but retains the concept of Turing [12]. **Quantum brain model**: Several proposals have been made for quantum brain models [37, 38] over the last decades and extended the range of quantum coherence [39] by microtubules.

Table 5.1 12 different pathways to create the time crystals, their editing capability, and location [12]

S.No	Possible elements	Edibility	Location
1	DNA, RNA, proteins	Edit it by changing seed, period	Period of periods
2	DNA, Microtubule, SNARE	Twist of twists, no of filaments	Spiral, the spiral of filaments
3	23 cavities of 26,000 protein database	The orientation of helices to make cavities	Tensor & vector of helices
4	Microtubule, Actin beta Spectrin network	Induced dipolar loop (continuum dipole model)	Lattice memory
5	Chemical transmission cycles	Change the diameter, phase, delay or starting points	Chemical-electronic-ionic
6	Ion channel density	Edit the pattern of density	Leak density of cavity
7	Assembly of neurons, cortical columns	Pitch, diameter and length of the spiral nodes	The geometry of spiral nodes
8	Assembly of cortical columns or neural superhighway	Divergence parameter and scale repeat memory	Vortex, or fractal memory
9	Spinal cord	Eight geometries from teardrop to ellipsoid & their pole	Golden nodes on the linear chain
10	Functional domains of Brain	Bond length, wiring length, and distribution	Phase of electromechanical
11	Primes non-linear density	Not editable, resides in the phase hyperspace	Hyperspace geometric
12	Dodecanion geometric constraints	Not editable a geometric nonphysical entity	Phase space duality

5.3 Evolution and Perception in the Human Brain: The Necessity to Emphasize the Use of Symmetries

The components of the dinosaur's brain are not identical to the human brain. The lower lobe remains nearly the same in both brains, while the anterior lobe of the human brain grows at an unusually high rate. Humans are better to learn anything (like science, art, culture, etc.) than other species. There has also been an argument over awareness in all living species. Here, we will follow the engineering path to find the answer [15]. The arrival of maximum complexity in the frontal lobe results from limitless research because it can reach the smallest, finest, and most distant. However, this result does not guarantee understanding human consciousness. To understand this, we need to break the limit, cooperate with the limit, and redefine the limit that is everyday work. More stress on the anterior lobe suggests that the human brain will look almost like an ellipsoid [12]. The brain develops vertically. Symmetries in the human brain and body are fundamental [40]. The size of the brain increases the length of the axon inside neurons. Longer wires have less noise to transmit signals but consume much energy.

5.3.1 *Perception of Primes in the Human Brain*

Prime numbers play an important role in long-term memory and short-term memory. Our sensory organs are always present in the order of prime numbers. For example, two eyes show prime two; the three major parts are in the cortex region (prime three), the bundles of five different nerves in the connectome model (prime five), seven films in cortical columns (prime seven), and seven retinal columns produce light (prime seven). The cellular microtubule system with thirteen proto-filaments occurs in the entire species on the planet (prime thirteen). The hippocampus consists of 17 different sections where 37 rings move the input and output signals of the brain. We have found 23 types of glial cells, oligo-dendrites, and branches in neurons from the literature data. The human nervous system consists of 31 spinal nerves, 12 cranial nerves, and 43 PNS (peripheral nervous system) that sense the environment. 1, 3, 5, 7, 19, 37… up to 47, fifteen primes are in the Brodmann's areas. Our decision comes out as a hexagonal form of 120,000 cortical columns. We created 120,000 cortical columns in which each column was filled with seven types of different organic gels and used them on a humanoid bot. The applications of prime numbers are endless. Nature has created PPM in the brain-body decision system so that all require very small communication to create a common vision. Communication is coded by PPM in any region of the universe or Brain [16].

The primes integer seven regulates the ability of the human brain to process information [41]. In the core of the microtubule, the water's particles assemble into the symmetry of the primes. In biology, the symmetry of primes is very high. It is found from small-scale (proteins) to large-scale (DNA). Higher-order tensors such as the octonion or dodecanian are used to improve the symmetry of its element as we move toward the larger size of the organ [42, 43]. Synthesis of symmetry is one of the keys to creating phase metrics that deliver suitable frequencies for precursor molecules synthesizing brain jelly to reverse engineer the brain. Symmetry has been viewed in terms of computation from the film of the 'cerebellum and hippocampus' to the protein dimer that produces triplets, similar to triple of resonance spectrum or PPM in the experiment [15]. For the last 2 million years, nature has created a perfect PPM in the human Brain [12].

5.3.2 *Hidden Properties in Resonant Bands of Biomaterials*

Resonant frequencies are not random; they appear in groups and are transformed into geometric forms such as triangles and rectangles, which reduces energy loss and forms a chain of vibrations [12]. A group of 3D geometric shapes can be observed throughout the configuration. This idea can be examined for the whole universe or the human brain because we find spirals everywhere. A genetic system is a guest time crystal that tries to match the universal chain of vibration or become its guest time crystal. PPM controls this process, but many PPM networks change from their infinite to infinite series. This change in a biological system is called evolution [44].

Most researchers ignore the geometry of the brain. For them, the copy of the circuits of neurons is everything. Here, we have studied all the structures but did not see the structure in terms of dimension. We have only noticed the symmetry of the constituent parts of the hippocampus, hypothalamus, proteins, glands, etc., at all spatial scales. Here we looked at the symmetry of the object. Symmetry is the lattice arrangement of system components, not their numbers. Here, the brain map is not just wiring of undefined regions, but a list of symmetries that create higher symmetry. However, we have filled the empty spaces in the brain with wires where the structure cannot vibrate. In a gold crystal, using symmetry, we can avoid billions of particles when the energy is supplied to the crystal. A plane of symmetry absorbs energy as a unit; however, different energy packets are not absorbed by different atoms. The absorbed energy is transferred from one active plane of symmetry to another active plane under permanent energy pumping. The system periodically switches between two symmetries.

Here, we have reported that the symmetry breaks continuously. The particular energy in a material is selected at a particular resonance frequency of the active plane. Most of the signals pass through the plane without interaction. Microtubule resonates in the 6-8 GHz frequency range show the quantum cloaking, and the DC resistance is reduced by 10^3 times at the same frequency [12, 25]. The EM resonance disappears the active plane to pass through the electron, the matter-wave tunneling through it. A system has many resonance frequencies. Electronic, magnetic, and mechanical resonances occur through electrons, quasi-particles, ions, photons. Each carrier chooses a plane of component arrangement for resonating with a particular frequency.

When the data runs in the phase prime metric, PPM [45], it determines the GML (geometric musical language) to obtain almost all the active regions in the useless data (Bandyopadhyay and Fujita 2021a, b, c). The pixels in the cube change rapidly; if it has three quick-changing points and three more quiet domains, it will look like a triangle.

Each sensor is designed in a geometric shape to detect the scream produced by singularity and quiet communication. It checks how the shape changes to a new geometric shape. In the journey of computing, geometric music changes by what was previously purely information. We do not get actual information. We check how the shape of geometry is changed by symmetry division. This change is free from human bias. Then we get to see how the change happens again. The triangle becomes a square, then pentagon, then a triangle again; this loop runs regularly. Then we draw a circle and, touching its boundary, draw a triangle, square, pentagon shape, etc.; this is the time crystal [16]. Our information is based on how the material breaks and forms the symmetry and creates a new symmetry in every structure.

A particular element in computing with primes is the phase prime metric, PPM, which expands the time crystals obtained from an event in nature over time. Integer space is a map of all possible interactions of prime. When an event is written as symmetry, there is a limited number of geometric symmetries. PPM can extract the solution of that event and show every possible solution.

This shows that past, present, and future architecture incorporate time crystals into themselves. PPM has several implications for the extension of an event in the form of symmetry breaking. We produce the information that does not exist, and we add those features in the future that had no past [12].

5.3.3 Concept of the Singularity Point

Singularity is defined by the gap in the phase space, where the phase structure of the biomaterial's signal transmission is undefined. Input does not matter for output. At such levels, the system vibrates, releases, or absorbs the signal at a particular frequency. There are many types of singularity, such as phase singularity, amplitude singularity, and polarization singularity. The singularity point is in a coiled element or field vortex at the corners of the geometric structure. The system point in the loop is a clock that passes through the shape's corners one by one. The signal explodes when a system passes through points of singularity [12]. Feynman excluded the singularity and 3-structure of paths in his reevaluation [46, 47].

5.4 How the Brain Uses Primes for Designing Its Components

Theoretically, the interaction of the electromagnetic wave with the biomaterial is detected by electromagnetic simulation software. To find the actual structural resonance peaks, we checked a wide frequency range (Hz–PHz) and detected resonances in this frequency range for all the components studied here. Suitable parametric simulation conditions and proper selection of properties of a material are necessary to find the actual resonance of biological material. All built dielectric materials are solved by the Maxwell equation solver in open boundary conditions and time-domain mode [48]. By pumping energy through waveguide sources on both sides of the cortical column and at the ends of dendrites and axonal branches of neurons, dielectric materials are stimulated. We scanned a wide frequency range to find the reflection and transmission coefficients.

5.4.1 Cerebellum

A fractal neural network includes thirteen cavities in the cerebellum; the seven layers are included in the cavity boundary. The entire structure has two major folds in the horizontal cross-section and seven major folds in the vertical cross-section view (see Fig. 5.1). Its structure looks like a network of trees of neurons, and it reads the final output from the hippocampus. Multiple forms of fractal wiring of neurons enable it to learn a complex range of conditions of mechanical movement. All sensors receive synchronized signals by the cerebellum to develop the decision-making process, similar to a device with high accuracy [49]. We are using frequency fractals as the origin of information processing. Frequency fractals are created by fusing the time crystals associated with all sensors. In this case, we do not need additional hardware. Which signal goes to which component of the brain can be understood from a simple chart, as reported in [50]. When the final decisions come from the hippocampus, the cerebellum controls the mechanical vibration that helps process memory. Dual clocks run [51] and depend on the duration of the signal emitted by the hippocampus. Suppose we connect the wires to the cerebellum. A correct signal will go to the correct output device.

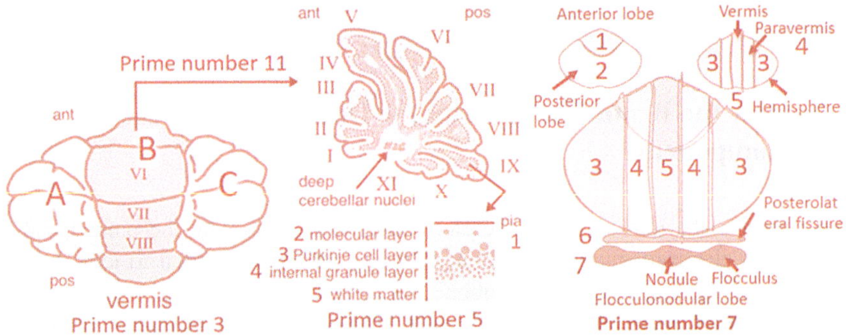

Fig. 5.1 The cerebellum geometry is present here in prime order. A schematic of the cerebellum is shown in the left panel. C3 symmetry (**A**, **B**, and **C**) is seen in three major regions (left lobule-A, vermis-B, and right lobule-C) of the cerebellum. Cerebellum middle part vermis (**B**) consists of 11 types of asymmetrical folding I, II, III, IV, V, VI, VII, VIII, IX, X, and XI. Each folding has five cell layers; pia-1, molecular layer-2, Purkinje cell layer-3, internal granule layer-4, and white matter-5 (mid panel). The major parts of the cerebellum include; anterior lobe-1, posterior lobe-2, left & right hemispherical region-3, left & right paravermis-4, vermis-5, posterolateral fissure-6, and nodule & flocculonodular lobe-7 or have the C7 symmetry

5.4.2 Hippocampus

Hippocampus consists of 17 cavities (Fig. 5.2, top right) with two spiral pathways running in parallel to each other (Fig. 5.2, bottom). Fivefold symmetry (Fig. 5.2, left middle) is found in such a pair of parallel pathways. The input signal coming from the whole body has two functions in the hippocampus. The first is to generate a temporary temporal code of the Brain [52] and second is to maintain its long-term memory [53]. A hippocampus has varying temporal codes for each type of memory [54], although chemical rhythms initiate long-term memory [55, 56]. First, a copy of the information [57] is affixed to the hippocampus. The information must be irradiated outside the upper brain to match the final information generated in the hippocampus [58] with the data already stored in it. The hippocampus would then

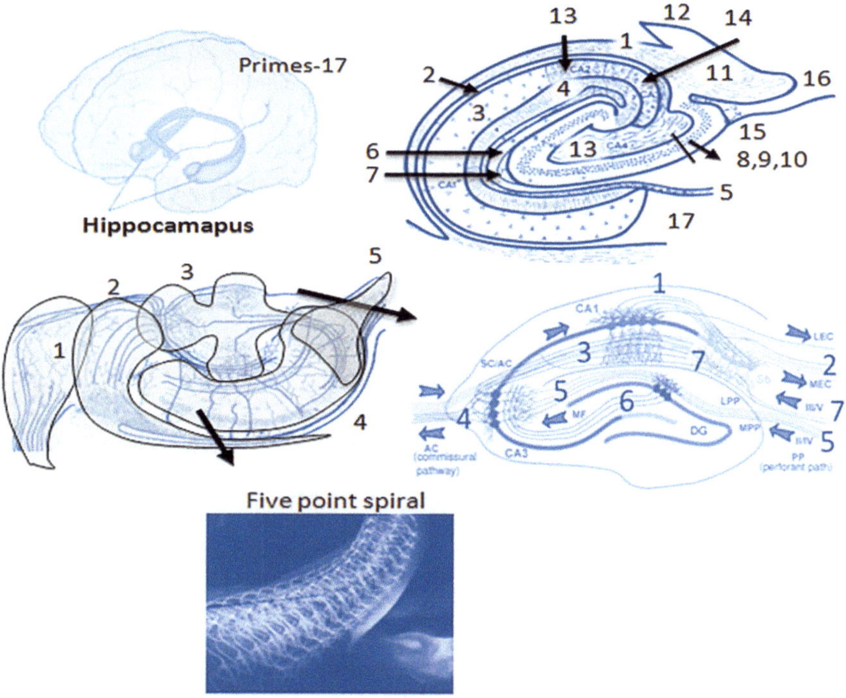

Fig. 5.2 Hippocampal geometry is arranged in prime numbers. A schematic of the hippocampus is shown in the top left panel. Five major regions 1, 2, 3, 4, and 5 are found in the hippocampus structure (middle left panel). Seven distinct nerve fiber bundles or layers exist inside those regions such as; CA1-1, MEC (medial entorhinal cortex)-2, SC/EC subicular or entorhinal cortex-3, EC (entorhinal cortex), commissural pathways-4, CAII/IV(MPP)medial pathways-5, MF Mossy fiber pathways)-6, CAIII/V (LPP) lateral pathways-7 (middle right panel). Each pathway is interlocked by five spiral rings (bottom left panel). A 2D cross-section view of the hippocampus is shown in the top right panel with 17 layers of nuclei (1–17)

function as an antenna receiver system. It is often activated at a selected frequency band [59].

5.4.3 *Hypothalamus*

The hypothalamus is a small part of the human brain used to generate rhythm and regulate the vital motor. The hypothalamus has 13 components or cavities. A schematic diagram of the hypothalamus is shown in Fig. 5.3, in which each component is shown by its name. The hypothalamus regulates the electromagnetic signal in the feed-forward bidirectional, which is evident by regulating metabolic clock, body temperature, emotional response, and physiological cycles [60]. These features are not enough for the device we want to build. If we want to make a high-level transformation in the final decision of the hippocampus, then we need to program those protocols here. Neurons self-assemble like periodic fractal seed generators, and an antenna composed of neurons transforms the final decision fractal within a certain

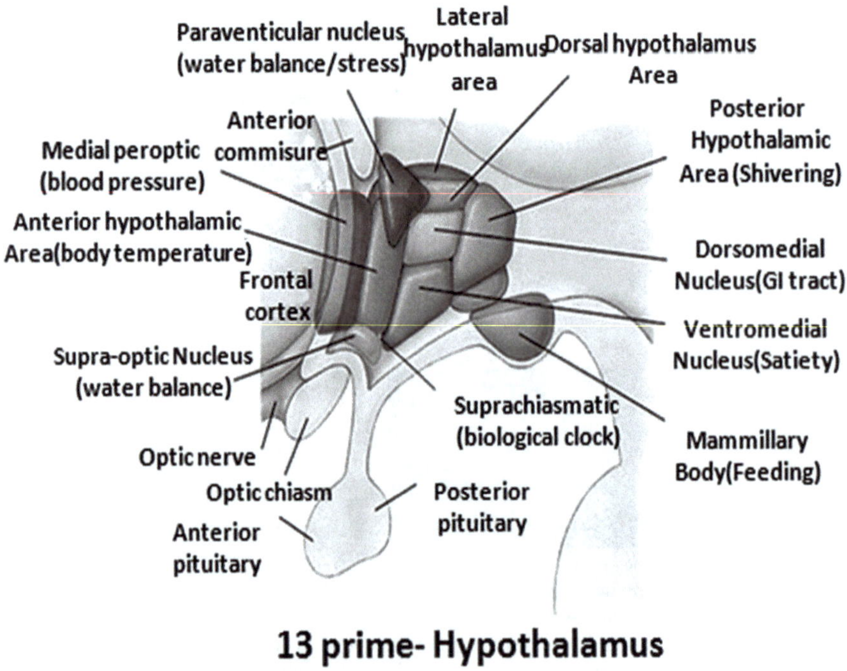

Fig. 5.3 A schematic diagram of the hypothalamus with 13 distinct nuclei; mammillary body-1, ventromedial nucleus-2, dorsomedial nucleus-3, posterior hypothalamic nucleus-4, dorsal hypothalamus nucleus-5, lateral hypothalamus nucleus-6, paraventricular nucleus-7, medial preoptic-8, anterior hypothalamic area-9, supraoptic nucleus-10, optic chiasm-11, pituitary-12, and suprachiasmatic nucleus-13

limit. The hypothalamus plays an important role in emotional functions such as love, pain, intimidation, anger, fear, and hate.

5.4.4 Connectome

The connectome is the map of fiber pathways in the Brain [61]. Learning about the properties of brain nerve fiber tracts is essential to understand the functional mechanisms of human brain architecture [62]. The function of the brain depends on the signals from the nerve fiber and their connection with regions of the Brain [63]. Psychiatric and neurological disorders arise by abnormalities in brain wiring [64]. Ya et al. [65] described the reconstruction method of brain fiber reconstruction and nerve fiber connectivity analysis using fiber orientation distribution tractography (FO-DT) [66]. Several techniques calculate fiber tract tractography based on diffusion, diffusion tensor magnetic resonance data (DT-MRI) [67], 3D polarized light imaging (3D-PLI), etc. [24, 68–71]. Those are the key to the conductivity of the brain nerves and understanding the continuity in fiber pathways [72, 73] have developed the algorithms for tractography based on the number of connections and synapses junction between all neurons for the reconstruction of nerve fiber pathways with a wide range of applications in neuroscience [74]. Furthermore, the geometrical map of the human connectome is important to understanding the functions of normal and pathological brain [75]. The intrinsic geometrical framework of the human connectome is sensitive to sensing changes in the connectivity of brain nerve fibers. Complex geometry with the high dimension of the brain connectome can be embedded in low dimension using a coupling pattern encoded technique to remove redundant variables and experimental noise [76, 65]. Robert et al. [77] investigated the geometrical role of network topology and described it by re-sampling brain fibers. Functions of integration and isolation are helpful to understand the relation between structure and function of connectome [78, 79]. The connectome is the nerve fiber pathway of the human brain and includes five distinct nerve pathways (see Fig. 5.4).

The human brain connectome is an interchangeable medium between wires and neuronal networks, so understanding wiring is essential for understanding the behavior of the brain [81, 82]. Previous studies have shown that neurons are placed inside the brain in a manner that is organized in such a way as to reduce wiring. In contrast, some studies showed that wiring relates to the topology of the connectome of a mouse. The topological arrangement of nerve fiber changes with age, whereas long nerve fiber bundles show a notable effect on brain functions [83]. Nerve fiber bundles or connectomes enhance our understanding of the brain [82]. Connectome receives information from pathways that travel to the cerebral cortex region. To find out the relation between the connectome and cortex layer, we need to understand the principle of nerve wiring that exists in the brain [84]. The topmost layer of the brain consists of cortical columns. Cortical columns are cells organized in a cluster form and can tune to resemble information coming from the brain nerve fibers [85]. The concept of the cortical column came from Mountcaste's Golgi studies [86]

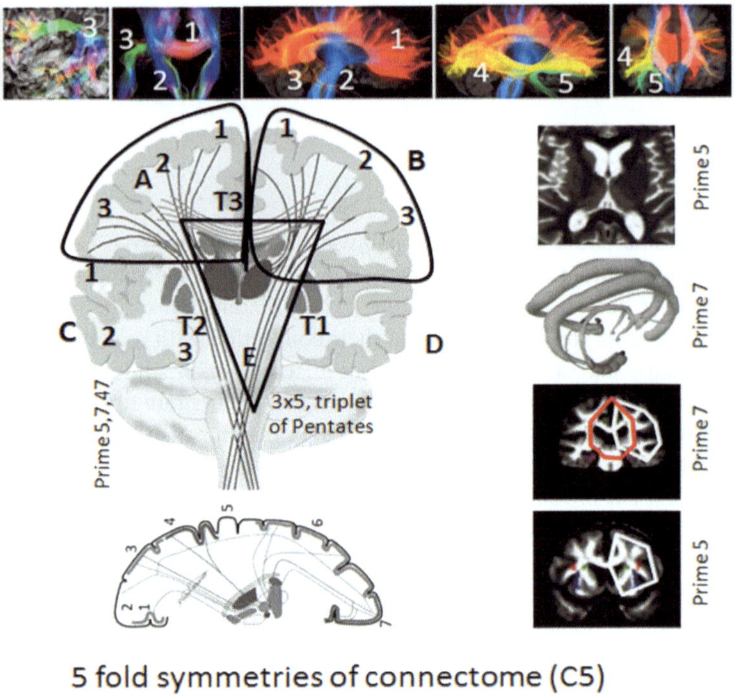

Fig. 5.4 There are five distinct pathways or C5 symmetries of nerve fibers in the connective map. (see top panel). During functional imaging of the connectome, we observed three distinct domains (T1, T2, and T3) activate five other domains or seven domains (right bottom panel). In the bottom panel, we reveal the sevenfold at cortex layer. Functional logic of fivefold and sevenfold established correlations [80]

through the visualization of the vertical chain of neurons [86] during cell mapping in the somatic cortex of the cat. How the column is the primary unit of the cortex is described in Philipp [87]. The column has 2 mm thickness and is defined by six input or output layers that are interconnected. Thousands of neurons inside the layers are functionally similar and show the response of the column [88]. The architecture of the column changes continuously in areas of the brain [89, 90] when it is activated. During the input and output process of the signal, columns try to maintain a normal pattern with columns present in other areas of the brain. A study provides evidence of a sub-network of neurons inside the cortical column. The layers produce magnetic fields in the cortical column in songlike cosmic structures [91].

The microtubule in the brain is the first-generation decision-making device. It is a hollow cylinder made of tubulin dimer (α and β) in which water is found in the spiral from inside and outside. The filament bundle or axon made of microtubule, actin, β spectrin, and other filaments arrives in the second generation of decision-making devices. At the same time, cortical columns composed of the structure of neurons, glial cells, and oligodendrocytes come in the third generation of decision-making

[12]. The cortical columns are arranged in the cortex layer of the brain as hexagonal. The entire layer consists of 47 groups of cortical columns, so Brodmann's area is covered by 47 columns [92, 93]. A feature of cortical columns is that a hexagonal packing can operate independently and produce a group reaction in a group [94]. Neurons within the cortical column find symmetric and asymmetric paths and float in different phases [95].

5.4.5 *Spinal Cord*

Thirty-one pairs of spinal cords converge in five distinct domains (C5) that transmit sensory signals from the entire body (see Fig. 5.5) to the brain. We can see spinal cord geometry in prime numbers. Suppose there are no firings in the spinal cord and medulla. In that case, there is a musical neuronal signal to balance the signal [96], indicating no continuous firing, clocking, or rhythm in the spinal cord [97].

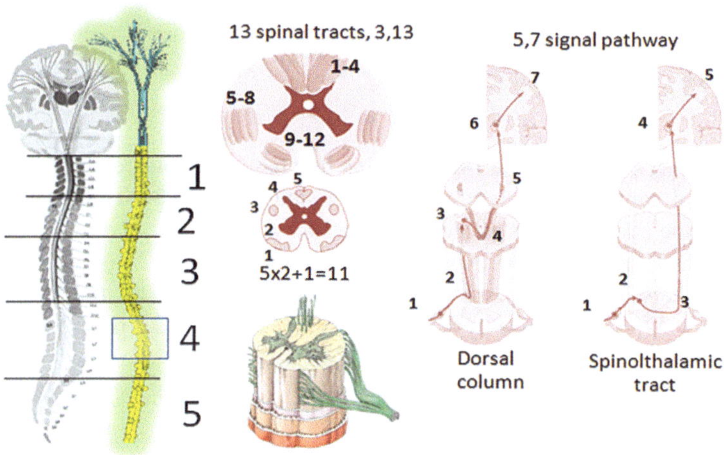

Fig. 5.5 A mimicked map of the spinal cord with its five sub-spinal regions; cervical (1), thoracic (2), lumber (3), sacral (4), and coccyx nerve's region (5). The spinal cord consists of 31 pairs of bones present in 5 regions. There are 11 regions inside the spinal cord. Out of it, 13 nerve bundles or spinal tracts pass, which collect the data from different parts of the human body and send it to the brain by 5 (dorsal) or7 (spinothalamic) signal pathways

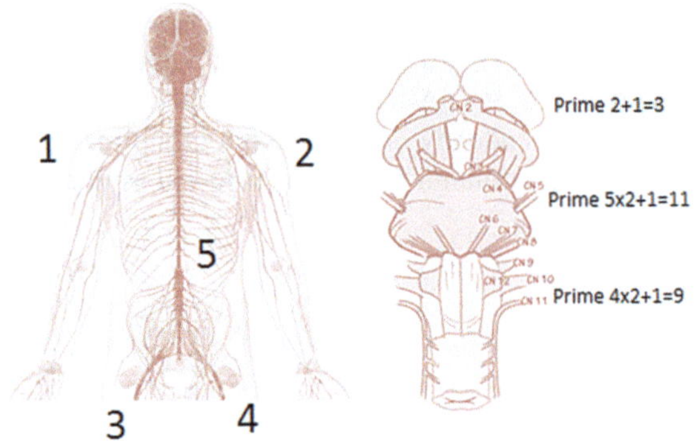

Fig. 5.6 The human nervous system consists of five major regions; 1—left-hand nervous system, 2—right-hand nervous system, 3—left leg nervous system, 4—right leg nervous system, and 5—stomach region nervous system or C5 symmetry. Cranial nerves in the human brain (olfactory nerve (limbic area)-I sensory, olfactory (eyes area)-II sensory, oculomotor nerve, trochlear and abducens nerves-V motor nerve, vestibulocochlear nerve-VI sensory, glossopharyngeal and hypoglossal nerves- VII and VIII motor nerve, vagus nerve- IX autonomic, sensory and motor nerve (mixed), spinal accessory nerve- X mixed, facial nerve- XI mixed, trigeminal nerve – XII sensory) show the left and right symmetry of C2 symmetry

5.4.6 *Cranial Nerve*

A higher level of time crystal fusion begins by phase coupling between local time crystals and cranial nerves (e.g., vagus with sinus node; [98]. Three cavities are present in the cranial nerves inside the midbrain.

In the first cavity = two cavities vibrate around a center, $2 + 1 = 3$ dynamic points; in the second cavity = 5 nerves ($5 \times 2 + 1 = 11$); in the third cavity = 4 nerves ($4 \times 2 + 1 = 9$); total dynamic centers $3 + 11 + 9 = 23$, (Fig. 5.6).

5.4.7 *Blood Vessel*

A blood vessel in the brain vibrates mechanically like a heartbeat. Using an ultrasound wave, we stimulated the blood vessel network (Fig. 5.7) by electromagnetic noise and vibrated the structure. However, the blood vessel network is connected to the

Fig. 5.7 A schematic of the blood vessel is shown here. It has five symmetric regions (1, 2, 3, 4, and 5) or C5 symmetry

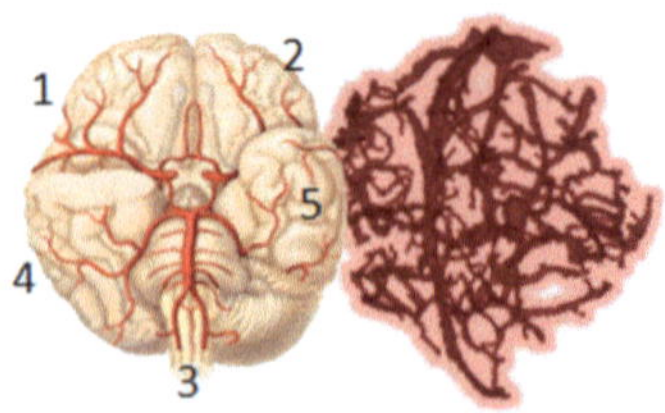

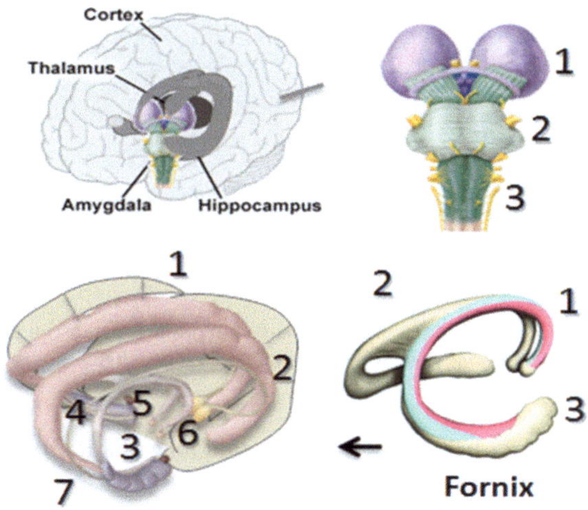

Fig. 5.8 A schematic of the limbic system is shown in the top left panel. The limbic system consists of cortex area, and the sets of central brain components; set1; thalamus, midbrain, and pons includes C3 symmetry (right-left panel) and set2; cingulate gyrus1, fornix2, amygdala3, hippocampus4, column of fornix 5, mamillary bodies 6, and amygdala seven or have C7 symmetry (bottom left panel)

connectome. Thus far, no studies have been done on how the blood vessel network affects the performance of the connectome.

5.4.8 Fornix

The C7 symmetry (Fig. 5.8, bottom left panel) is found in fornix geometry, acting as a transmitter and receiver antenna between the upper brain and midbrain [99]. It has the triangular form 1, 2, and 3 (Fig. 5.8, bottom right panel). The hippocampus and neocortex have a learning process that performs in a complementary pathway [100]. The cingulate gyrus, positioned inside the fornix, acts as an antenna for the brain to learn. The interaction between cortex and hippocampus indexes memories and forms the time crystal [101].

5.4.9 Brodmann's Region: Upper Cortex Layer

The map of cortex region is based on shape of neurons and other nerve cells, and how cortex layers are viewed. Different brain regions or Brodmann's regions are identified from 1 to 47.47 brain regions are identified by their names and functions as follows: 1, 2, 3, 5, 7, 31, 39, 40 parietal lobe (body sensation); 4, 6, 8, 9, 10, 44, 45, 46, 47 frontal lobe (motor); 11,12,24,25,32,33,38 anterior cingulated and orbital cortex (emotion); 17,18,19,20,21,37,38 occipital cortex and temporal cortex (vision); 23, 26, 27, 29, 30, 35, 36 medial temporal lobe, posterior cingulate cortex (memory); 22, 38, 41, 42 temporal lobe (audition); 28, 34 medial temporal cortex

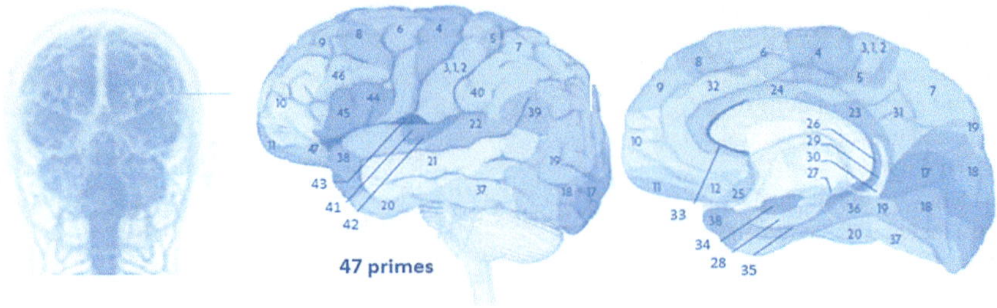

Fig. 5.9 The cortex layer is arranged in prime order, and it has C47 symmetry. The outer and 2D cross-sectional view of the cortex layer is shown here

(olfaction); 22, 38, 41, 42 temporal lobe (audition). The cortex layer of human brain has 47 primes (see Fig. 5.9).

5.4.10 *Thalamus*

Thalamus is a gateway to the sensors, cerebral cortex [102], and the cerebellum. It forms 13 cavities or possesses C13 symmetry (Fig. 5.10) and behaves as the universal synchronizer [103] that maintains the natural logarithmic relationship between brain

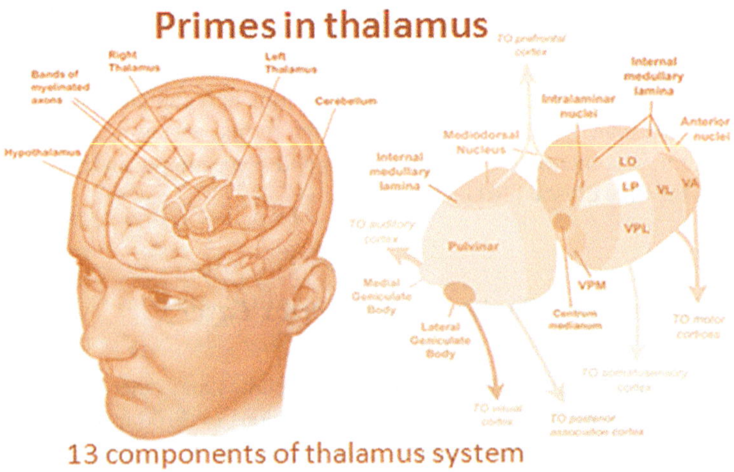

Fig. 5.10 Inside the human brain, two ellipsoid-shaped nuclei are arranged side-by-side. The thalamus is a relay station that senses the signal coming from different body parts and sends it to the cerebral cortex. The thalamus model has 13 nuclei, midline nuclei, intralaminar nuclei, anterior nucleus, lateral dorsal nucleus, lateral anterior nucleus, lateral ventral nucleus, reticular nucleus, lateral posterior nucleus, lateral lamina, medial ventral posterior nucleus, centromedian nucleus, medial dorsal nucleus, geniculate nucleus

components [104]. The system looks as if many springs are side-by-side; if one starts vibrating, a complex vibration occurs in the entire system. Each spring accepts specific types of sensory information [105]. Thalamus works along with basal ganglia [106].

5.4.11 *Cortical Column*

A schematic diagram of a cortical column is given in Fig. 5.11. The cortical column is the memorizing fundamental component that looks to be a more complex operation than the individual neuron. A common feature of a cortical column is the organization of vertical seven layers in it, leading to C7 symmetry. Different layers of neurons in each layer occur in different compositions and form different connections with

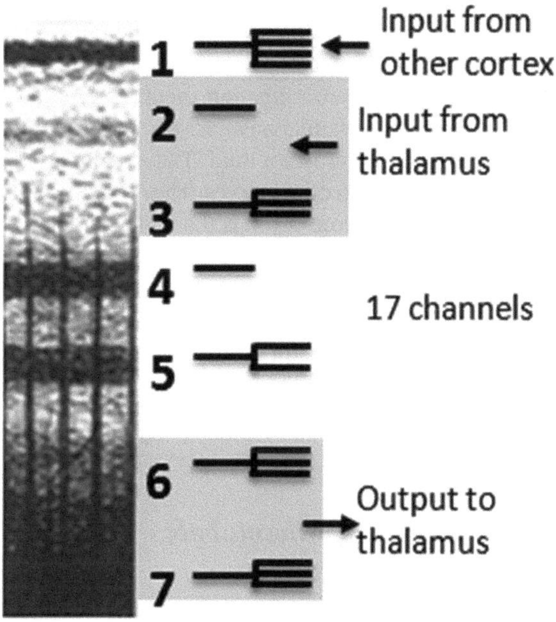

Fig. 5.11 There are seven layers in the cortical column. Layers 6–7 have different functions from upper layers 1–3. The bottom layers serve to send and receive signals from other cortical columns and cortex regions. The signals from upper layer 1–3 go into layers 6–7 and send them to the bottom layer instead of responding to the thalamus and other sensory regions. The thalamus sends the information from the cortex and other subcortical areas into 4–5 layers, dividing the column into upper and lower layers. Mid-layers 4–5 form a feedback circuit by passing signals through the thalamus and return to layers 4–5 with 1–3 layers. The thalamus integrates additional internal and external information. Layer 1–3 and layer 6–7 act as input and output channels for signal communication with other cortical columns due to their long width [107, 108].

nearby neurons. Despite neurons in different layers sticking to other neurons, such an architecture is called a key cortical column [85].

5.4.12 Neuron

The neuron is the major component of the brain that interacts with chemical and electrical signals. Signal communication between neurons occurs by synapse junctions. There are billions of cells in the human body. The nervous system of our body consists of specialized cells. Neurons hold the nerve impulse. The neuron is a fundamental element of information processing. Neurons transmit the information as electromagnetic signals, which find necessary changes required in the hardware. Accordingly, the synapses fire. There are two types of cells in our nervous system: glial cells and neurons—Glial cells are non-conductive cells that are important for structural support. In contrast, neurons are conductors that transmit information from one place to another. There are 100 billion neurons in our brains. Neurons have many parts, such as the plasma membrane, cytoplasm, axon, soma, and dendrite that play an important role in playing neurons' function (neurons type and function). The behavior of the plasma membrane is like a barrier to cytoplasm inside neurons, but it excretes some substance that flows like a fluid. The structure of dendrites is like a tree shape that receives signals from other nerves. The dendritic branches of neurons carry signals to the soma (cell). **Axon**: Axon is the main conductive unit of neurons and has a role in cellular communication. The axon is a long and thin branch that carries signals away from the cell body. **Soma**: The soma is the spherical part of neurons in which the nucleus is contained. The soma or cell body receives signals from the axon terminals and transmits this signal to axon branches that depend on the signal strength. Here 26 classes of the geometry of neurons are described. The different types of neurons shown in Fig. 5.12 have different resonance frequencies.

5.4.13 Nerve Filaments and Microtubule

Nerve fibers are not cable-like. The periodic oscillation of the signal transmitted from neurons means that it reflects the architecture of the clock inside the clocks [109]. There are various types of nano-viruses inside nerve fibers, such as neurofilament (10–15 nm wide), actin (2–4 nm wide), and microtubules (25–30 nm wide). Thousands of such neurofilaments are silent from previous decades, while they play a role in firing neurons and transferring the signals [110, 111]. By selecting a variable pitch of spiral, one can produce different frequency signals in the case of microtubules (Fig. 5.13). In short, microtubules are capable of holding a wide range of primes for resonance.

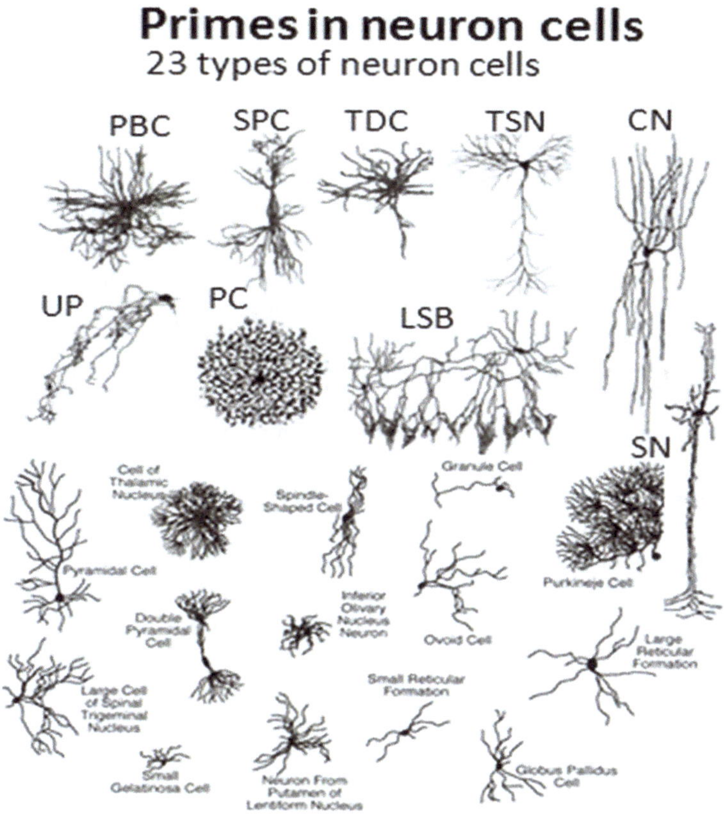

Fig. 5.12 A photograph of 23 types of neurons

Fig. 5.13 Microtubule geometry in prime order

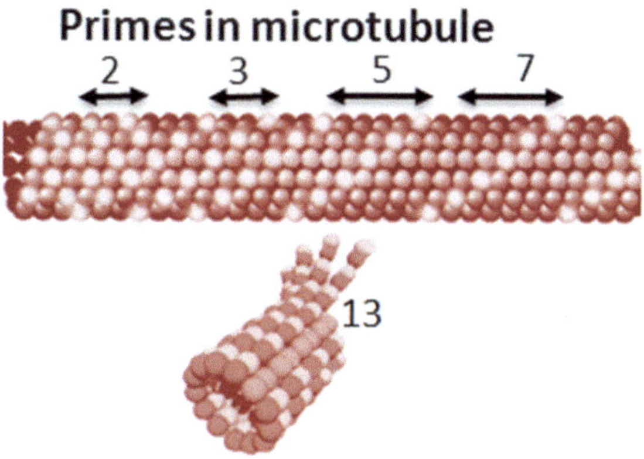

5.5 Simulation Study of the Brain-Body System of Human

5.5.1 Cerebellum

Four sensory probes or ports are placed at the location shown in Fig. 5.14a, top part. The simulation ran for several days to optimize the bottom part's reflection and transmission spectra obtained in Fig. 5.14a. The spectrum between each pair of ports is noted in KHz frequency domains. Solving the Maxwell equations of large numbers of fibers are requires extreme patience in placing the probes at the right location around the 3D architecture of the cerebellum. A little change in the location and area of the port would lead to the divergence. Four waveguide sources, ports 1, 2, 3, and 4, are placed at inferior vermis, superior vermis, left horizontal fissure, and right horizontal fissure, respectively. Four sharp resonance peaks are identified by the shown curve at the bottom of Fig. 5.14a. At 250.70 kHz, the quality factor is the highest. Hence we imaged electric and magnetic field distribution at this particular frequency, as shown in Fig. 5.14b. One could conclude that the identity of ports does not reflect in E-B distribution. From an intensity scale, we have shown that particular parts that eventually connect to the midbrain, nerve fibers in the cortex are naturally selected as a part of the resonant electromagnetic oscillations.

Simulation details: Used solver—Maxwell equation solver; selected mode—time domain; boundary condition—open space; waveguide port dimension—40 mm × 60 mm; frequency domain 140–175 kHz range.

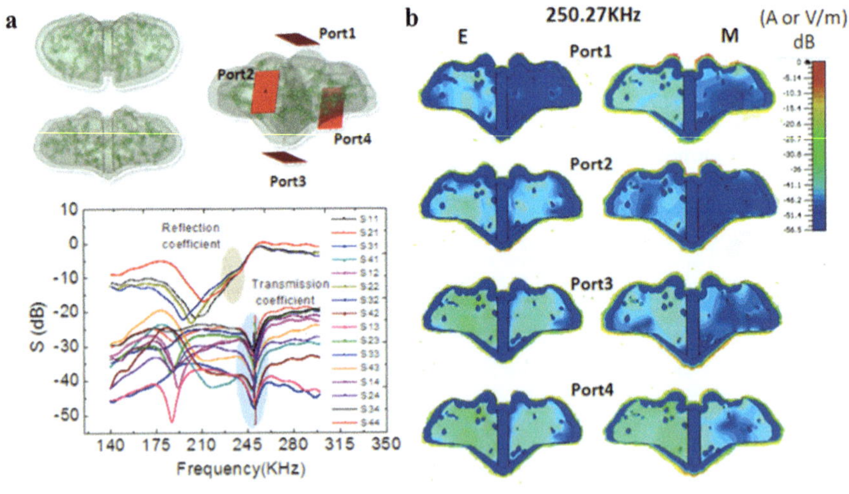

Fig. 5.14 **a** Constructed cerebellum model and simulated resonance band are shown in the top and bottom panels, respectively. **b** EM field distribution on surface and inside of cerebellum model

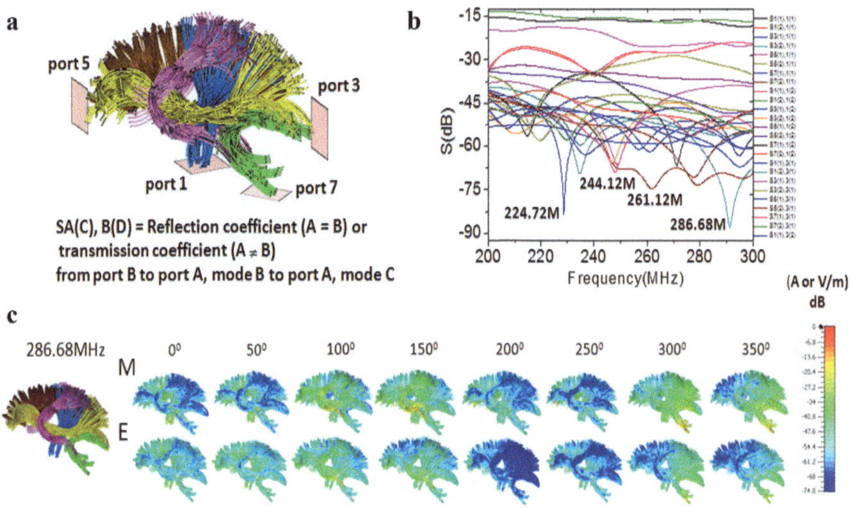

Fig. 5.15 **a** This panel shows the constructed cerebellum model. Four waveguide ports are connected to different nerve bundles. **b** Simulated resonance spectrum for all waveguide ports. **c** EM field distribution on connectome surface is observed at the resonance frequency (286.68 MHz) over one complete phase cycle. M and E are represented as magnetic and electric fields

5.5.2 Human Connectome Model

The human connectome model was built by considering its original dimensions and solved in Maxwell equation solver (time-domain mode) by placing four waveguide energy sources in its various regions (Fig. 5.15a). The resonant frequency spectrum for each power source is seen in the MHz frequency range with different resonance peaks (Fig. 5.15b). The EM field distribution profile of the connectome model of the human brain is observed at a 219.6 MHz resonance peak. Both fields show the periodical nature with maximum (100°–150°, 300°–350°) and minimum (0°–50°, 200°–250°) intensity at the parietal lobe (Fig. 5.15c).

Parametric details: Used solver—Maxwell equation solver, mode—time domain, boundary condition—open space, selected frequency range 200–300 MHz, applied waveguide port dimension 200 mm × 100 mm.

5.5.3 Cortical Column

A cortical column assembly consists of 19 cortical columns (Fig. 5.16, left panel). We stimulated the structure by applying the energy source at both ends of the structure. Resonance of built cortical column assembly is shown in the kHz frequency range, as shown in Fig. 5.16 (middle panel). We have obtained the profile of the electric and magnetic fields at resonance peaks 112.89, 266.33, and 268.12 kHz. In Fig. 5.16

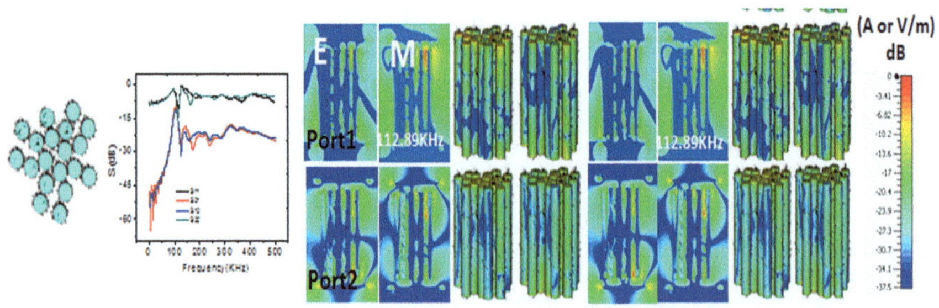

Fig. 5.16 The left panel shows artificial cortical column assembly. The middle panel is about the simulated resonance spectrum, and the right panel presents EM field distribution on the surface and a 2D cross-sectional view of cortical column assembly

(right panel), electric and magnetic fields show the dominating nature near the energy waveguide source and are distributed as a cluster at 112.89, 118.63, and 172.23 kHz resonance peaks.

Simulation details: Used solver—Maxwell equation solver; selected mode—time domain; boundary condition—open space; waveguide port dimension—25 μm × 25 μm; frequency domain—kHz.

5.5.4 Neuron

Simulated structures of four individual neurons (pyramidal neuron, small axon cell, chandeliers cell, and descending axon cell) in terms of resonance spectrum, electric and magnetic field distribution are shown in this Fig. 5.17. Two energy sources, ports 1 and 2, apply at the upper and lower ends of neurons, but we have only shown the resonance spectrums and EM field distribution profiles for port1. EM field distribution is measured at the resonance frequency of neurons. Resonance frequencies of all four neurons are 150.1 MHz-pyramidal neuron, 41.34 MHz-small

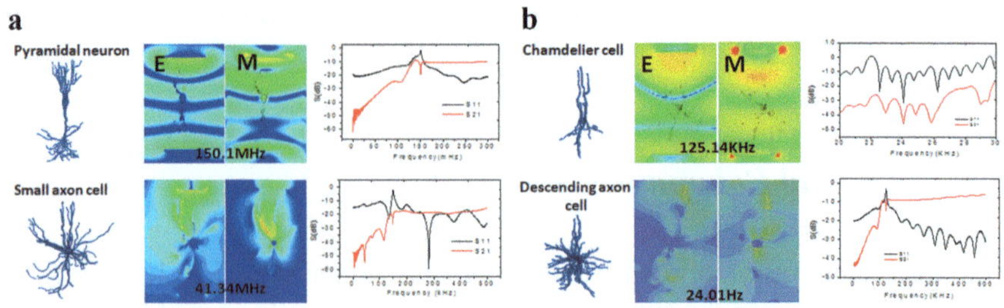

Fig. 5.17 Simulated resonance spectrums and EM field profiles at the resonance frequency of four types of neurons are shown here

axon cell, 125.14 MHz-chandelier cell, 24.01 MHz-descending axon cell. E and M are the symbols of the electric and magnetic fields, respectively.

Simulation details: Used solver—Maxwell equation solver; selected mode—time domain; boundary condition—open space; waveguide port dimension—2.3 mm × 2.3 μm; frequency domain Hz–MHz range.

5.5.5 *Hypothalamus*

Figure 5.18a (left) shows the simulation model of the hypothalamus. Resonance spectrums with reflection & transmission curves have been observed in the 400–700 kHz frequency domain by applying four waveguide ports; ports 1, 2, 3, and 4 at the mamillothalamic tract, fornix, suprachiasmatic nucleus, and optic chiasm, respectively (Fig. 5.18b, left). EM field distribution profiles at 497.63 kHz resonance peak (Fig. 5.18b, right) are shown in the right of the panel for all applied energy sources; port 1, 2, 3, and 4 (Fig. 5.18a, right).

Simulation details: Used solver—Maxwell equation solver; selected mode—time domain; boundary condition—open space; waveguide port dimension—300 mm × 300 mm; frequency domain 450–650 kHz range.

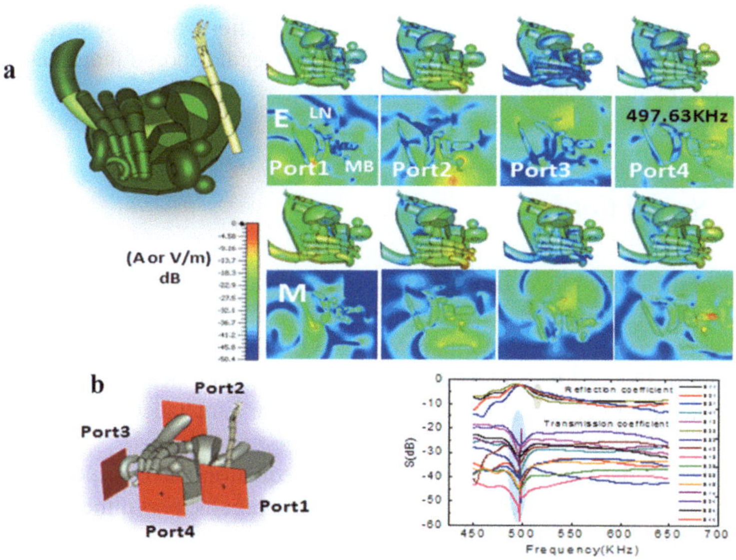

Fig. 5.18 **a** The left panel presents an artificial hypothalamus structure, and the right panel shows the EM field distribution at 497.63 kHz frequency. **b** Simulated resonance spectrum for four ports applied to different regions of the hypothalamus

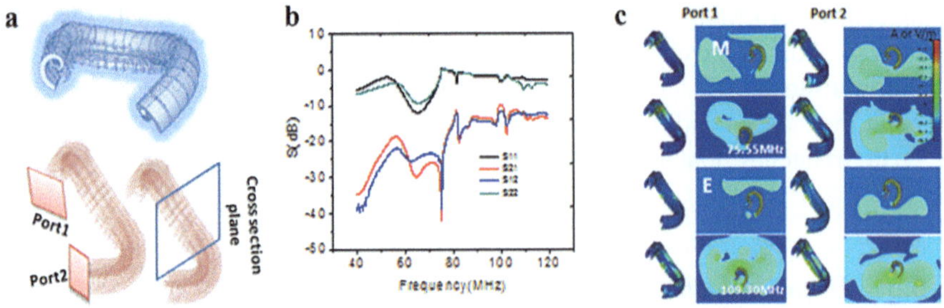

Fig. 5.19 **a** In hippocampal geometry, the hippocampus is stimulated by two ports put on both of its ends. **b** Simulated resonance spectrum. **c** EM field distribution at resonance frequencies on the outer surface and 2D cross-section plane of the hippocampus

5.5.6 Hippocampus

The mimicked geometry of the hippocampus is created in Maxwell equation solver a with port 1 and 2. Ports are applied at both ends, shown at the top of Fig. 5.19a, while the simulated electric and magnetic fields in Fig. 5.19c are shown in 2D cross-sections of the plane at 49.83, 47.71 MHz resonance peaks. Reflection (S11 & S22) and transmission (S12 & S21) peaks have been obtained in MHz frequency range (40–100 MHz) (Fig. 5.19b).

Simulation details: Used solver—Maxwell equation solver; selected mode—time domain; boundary condition—open space; waveguide port dimension—10 mm × 10 mm; frequency domain 40–100 MHz range.

5.5.7 Spinal Cord

Mimicked dielectric model of the spinal cord consists of 5 sets of spinal nerves, as shown in Fig. 5.20a. The built model is excited by placing port1 and 2 at both ends of its and resonance spectrum with peaks 1253.6, 1297.1, 1473.1, and 1495.1 MHz are shown in Fig. 5.20b. Simulated electric and magnetic field profiles are shown at resonance peaks (Fig. 5.20c) when both ports are simultaneously placed on the model.

Simulation details: Used solver—Maxwell equation solver; selected mode—time domain; boundary condition—open space; waveguide port dimension—10 cm × 10 cm; frequency domain—MHz range.

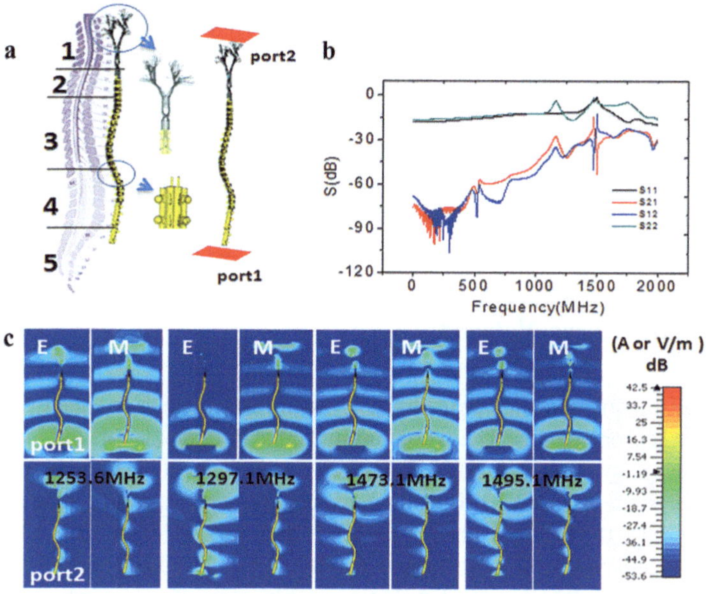

Fig. 5.20 **a** Built spinal cord geometry with two waveguide ports is applied at both ends of it. **b** Simulated resonance band. **c** EM field distribution for both ports at different resonance frequencies. EM field scale is shown on the right side

5.5.8 *Thalamus*

The thalamus model is created by considering all biological details. 4 waveguide ports; ports 1, 2, 3, and 4, are applied at the left-& right of both 'medial dorsal nucleus (port 1 and port 2) and 'lateral and medial ventral posterior nuclei (port 3, and port 4') of the thalamus and solved the structure in numerical simulation software (Fig. 5.21a,

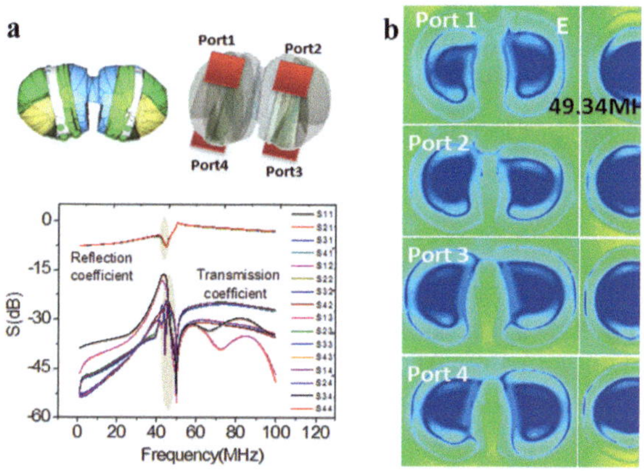

Fig. 5.21 **a** Built thalamus model with four waveguide ports. **b** Simulated resonance band. **c** EM field distribution for all waveguide ports at 49.34 MHz resonance frequency

top). A simulated resonance spectrum in reflection and transmission coefficients has been observed in the 0–120 MHz frequency range at 49.34 MHz resonance peak (Fig. 5.21a, bottom). Figure 5.21 b shows the corresponding electric and magnetic fields are estimated at 49.34 MHz frequency for all applied ports; port 1, 2, 3, and 4.

Simulation details: Used solver—Maxwell equation solver; selected mode—time domain; boundary condition—open space; waveguide port dimension—300 mm × 140 mm; frequency domain 0–100 MHz range.

5.5.9 Skin Nerve Network

A complex and massive dielectric model of human skin fibers is created by following all packed sensory nerves. The whole mimicked skin model is excited by pumping the EM energy by a suitable dimension of waveguide ports at the bottom and top side nerve net system (Fig. 5.22a). The simulated electromagnetic field for the individual port; port 1 (570 kHz, 933 kHz, 1.29 MHz, and 1.59 MHz), port 2 (2.45 MHz, 2.68 MHz), and combined ports; port 1, 2 (789.2 kHz, 1.15 kHz, and 1.78 kHz), are shown. Respective 3D EM field distribution profiles along the skin nerve net detected at resonance peaks of port 1 (570 kHz, 933 kHz, 1.29 MHz, and 1.59 MHz), port 2 (2.45 MHz, 2.68 MHz), and combined ports; port 1, 2 (789.2 kHz, 1.15 kHz, and 1.78 kHz) are shown in Fig. 5.22b. Here, E and M are electric and magnetic fields, respectively.

Simulation details: Used solver—Maxwell equation solver; selected mode—time domain; boundary condition—open space; waveguide port dimension—16 cm x 16 cm; frequency domain kHz–MHz frequency range.

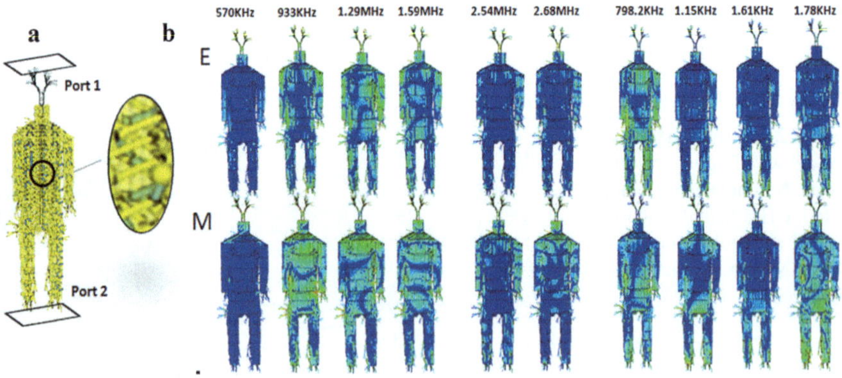

Fig. 5.22 **a** Built skin network, two waveguide ports are applied at top and bottom of it. **b** EM field distribution on the skin surface at different resonance frequencies

5.5.10 *Blood Vessel Network*

5.5.10.1 Mechanical Vibration of Blood Vessel: Connectome Functional Responses

For the last few decades, numerous studies have reported on the mechanical vibration on the physiological activity of humans and animals. The first time, Nakamura et al. 1996 suggest that vibration is the key can could have importance in research purposes. Low-frequency vibrations influence blood circulation [112, 113]. Such vibrations may be important to see the effects of high-frequency vibration on the human body [114]. Further improvement has been seen in blood circulation and chronic pain in the lower part of the spinal cord [115], blood pressure in the femoral artery, and other parameters regarding the circulatory vibration [116] when people are exposed by mechanical vibrations [117]. Blood flow in an effective area [118]. The effect of the mechanical vibrations on the blood flow, blood vessel size, and diastolic pressure in the low frequency have been reported in the literature [119, 120]. The blood flow rate increased by 33% during the vibration [121]. Different types and numbers of vibrations increase during treatment; however, several researchers studied whole body vibration to examine the relation between static and dynamic vibrations.

The mechanoreceptors (vibration-sensitive receptors) in our skin trigger during whole-body vibrations [122]. Such mechanoreceptors transmit a signal to the somatic sensory cortex region [122]. The prefrontal cortex region is linked with the sensory cortex regions by direct and indirect pathways involved with the cognitive processing regions [123]. The indirect pathway contains the limbic system, influencing the hippocampus's memory and learning process and amygdala in the prefrontal sensory cortex [124]. Emotion influences memory and the learning process [125].

5.5.10.2 How Does the Brain Beat with the Heartbeat?

In the brain, neurons fire, and information transmits. For the last few decades, researchers claim the neuron's membrane does everything. Inside the neuron, an axon integrates the structure of actin, beta spectrin, ankyrin, and microtubule. The previous study claims that the microtubule offers cognitive features. However, recent studies suggest that all constitutes of microtubule like alpha & beta-tubulin, beta plate contributes equally role in consciousness or conscious emerges at the deeper level in the microtubule during vibration. Cells/neurons are closely packed with tiny filaments, which provide strength to cells. Every cell in the human body vibrates with a certain frequency range, and they communicate with each other at a particular frequency. In our previous study, we have mapped 537 different types of rhythms in the human brain, which are interlocked by phase and time. All those rhythms offer 20 types of well-known consciousness features. The built-in complex circuitry model offers all features in the HBS brain, and EEG checks functional activities.

The essential information is played as a permanent memory in the brain by allosteric waves between various proteins.

The human body generates mechanical vibration at the low resonance frequency, produced by heartbeat and blood flow in the vessel. Every human body's component vibrates at a particular resonance frequency. Blood circulation has resonance frequency at 0.05–0.3 Hz. The heart vibrates at 1 Hz, while brain vibration occurs at 10 Hz. Such vibrations are linked with the cardiovascular system, which simultaneously occurs with the heart mechanism. Such vibrations are used to identify the emotional state by using the amplitude-frequency response of these vibrations. Literatures studies are based either on protein or whole-body vibration but not a single study reported on mechanical vibration of the blood vessel.

5.6 Construction of 20 Consciousness Circuits in Humanoid Bot' Brain

Brain circuits of 20 cognitive responses built inside the brain of humanoid bot subject HBS are shown in Figs. 5.23, 5.24, and 5.25. All clocks and loops are mapped from the reported studies [16]. Symbol details of brain components used in Figs. 5.23, 5.24, and 5.25 are given as.

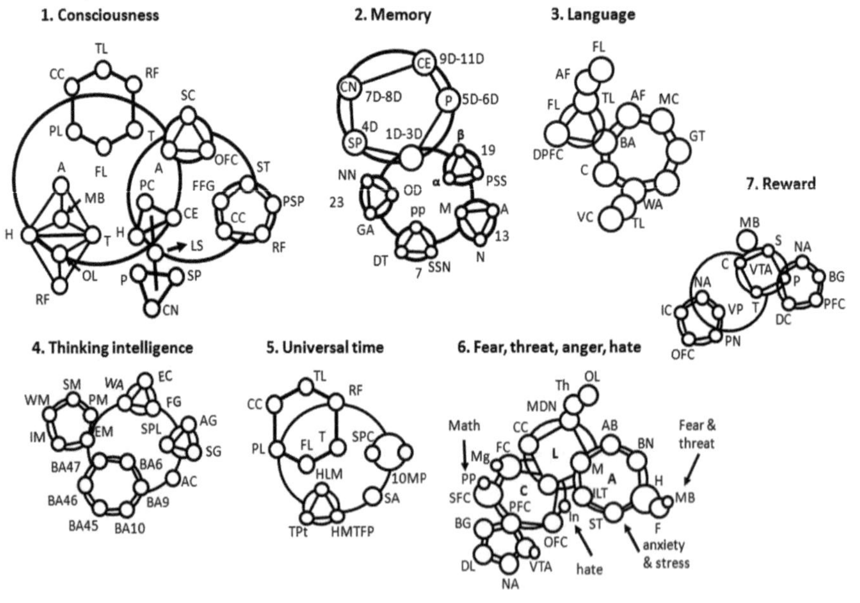

Fig. 5.23 Clocking map of conscious circuits used in humanoid bot brains—consciousness; memory; language; thinking and intelligence; universal time; fear, threat, anger, hate

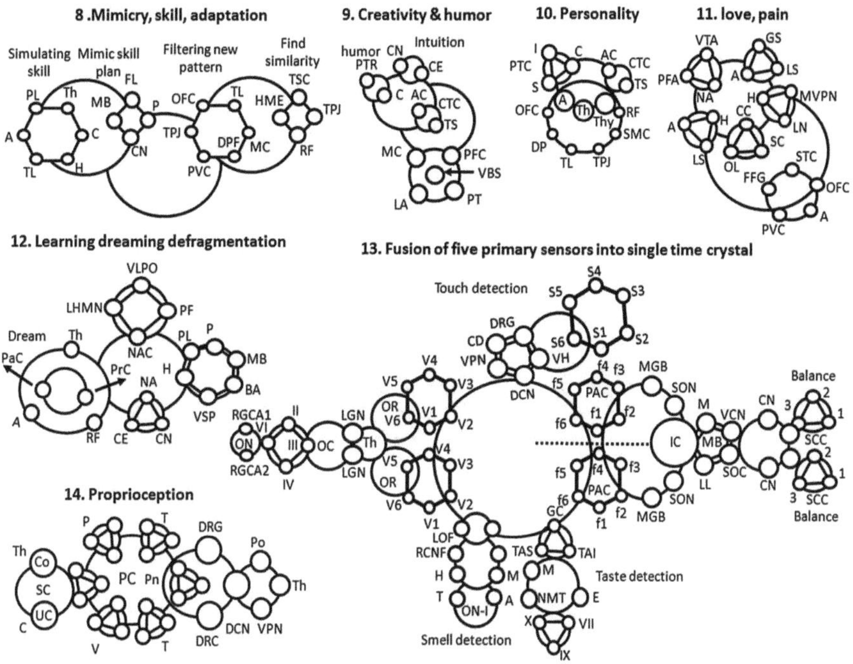

Fig. 5.24 Clocking model of brain's conscious state generation circuits: Mimicry, skill, adaptation; Creativity & Humor; Personality; Love and pain; Learning dreaming defragmentation; Fusion of elementary sensor into single time crystal

5.6.1 *Consciousness*

Central Executive, CE; Olfactory lobe, OL; Mamillary body, MB; Parietal lobe, PL; Corpus Callosum, CC; Temporal lobe, TL; Fusiform gyrus, FFG; Corpus callosum, CC; Primary sensory pad, PSP; Superior temporal, ST; Sensory pads, SP; Caudate nucleus, CN; Putamen, P; Amygdala, A; Hippocampus, H; Reticular formation, RF; Thalamus, T; Limbic system, LS; Superior caliculus, SC; orbitofrontal cortex, OFC; Frontal lobe, FL; Parietal cortex, PC.

5.6.2 *Memory*

Protein secondary structure, PSS; α-helix, α; β-sheet, β; Microtubule, M; Actin, A; Neurofilament, N; Neuron, NN; Glia-Astrocyte, GA; Oligo-dendrocyte, OD; Pyramidal & polar, PP; Tear-drop transform, TDT; Superstructures of neuron, SSN.

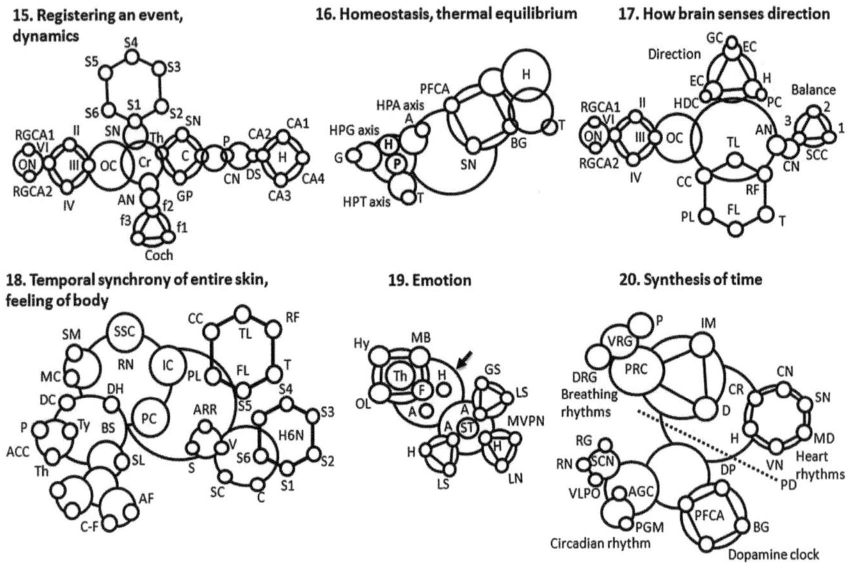

Fig. 5.25 Clocking map of conscious circuits-Registering an event dynamic; homeostasis, thermal equilibrium; How brain senses direction; Temporal synchrony of entire skin cover, the feeling of body; emotion; synthesis of time [12, 16]

5.6.3 Language and Conversation

Arcuate fasciculus, AF; Temporal lobe, TL; Frontal lobe, FL; Dorsal Prefrontal cortex, DPFC; Motor cortex, MC; Broca's area, BA; Gesdiwind territory, GT; Cerebellum, C; Wernicke's area, WA; Visual cortex, VC.

5.6.4 Thinking and Intelligence

Extrastriate cortex, EC; Fusiform gyrus, FG; Wernicke's area, WA; angular gyrus, AG; supramarginal gyrus, SG; superior parietal lobule, SPL; Anterior Cingulate, AC; Working memory, WM; Procedural memory, PM; Episodic memory, EM; Semantic memory, SM; Implicit memory, IM.

5.6.5 Sense of Universal Time, Symmetry

10 Metrics of prime, 10MP; Statistical prime contribution, SPC; Striatum amygdala, SA; Temporal lobe, TL; Thalamus, T; Parietal lobe, PL; Hippocampus long memory,

HLM; Topological projected time, TPt; Frontal lobe, FL; Corpus callosum, CC; Reticular formation, RF; Hippocampus min time filtered pattern, HMTFP.

5.6.6 Fear, Threat, Anger, Hate

Medical dorsal nucleus, MDN; Limbic system, LS; Thalamus, Th; Posterior portion, PP; Accessory basal, AB; Middle gyrus, Mg; Basal nucleus, BN; Cingulate cortex, CC; Cortex, C; Medical, M; Amygdala, A; Hypothalamus, H; Input lateral terminal, ILT; Olfactory lobe, OL; Fornix, F; Mammillary body, MB; Basal ganglia, BG; Frontal cortex, FC; Superior- Frontal cortex, SFC; Prefrontal cortex, PFC; Ventral tangential array, VTA; Dopamine level, DL; Styria terminalis, ST; Nucleus acumens, NA; Orbitofrontal cortex, OC; Insula, In.

5.6.7 Reward

Basal ganglia, BG; Nucleus accumbens, NA; Ventral tegmental area, VTA; Midbrain, MB; Prefrontal cortex, PFC; Dopamine cycle, DC; Striatum, S; Pallidum, P; Thalamus, T; Cortex, C; Ventral pallidum, VP; parabrachial nucleus, PN; orbitofrontal cortex OFC; and insular cortex, IC.

5.6.8 Mimicry, Skill, Adaptation

The parietal lobe, PL; Thalamus, Th; Cerebellum, C; Hippocampus, H; Amygdala, A; Temporal lobe, TL; Caudate nucleus, CN; Mammillary body, MB; Putamen, P; Frontal lobe, FL; Primary visual cortex, PVC; Temporal lobe, TL; Dorsolateral Prefrontal, DPF; Orbitofrontal cortex, OFC; Motor cortex, MC; Reticular formation, RF; Supplementary Motor cortex, SMC; Thalamus sensor cross-over, TSC; Tempoparietal junction, TPJ; Hippocampal memory encoding, HME.

5.6.9 Creativity and Humor

ACC Anterior Cingulate, AC; Temporal sulcus, TS; Caudate nucleus, CN; Central executive, CE; Left Amygdala, LA; Prefrontal cortex, PFC; Motor cortex, MC; Posterior Temporal region, PTR; Motor cortex, MC; Ventral brain stem, VBS; Cerebellum, C; Pyramidal tract, PT.

5.6.10 Personality

Supplementary motor cortex, SMC; Creativity time crystal, CTC; Personality time crystal, PTC; Insula, I; Cingulate, C; Striatum, S; ACC Anterior Cingulate, AC; Temporal sulcus, TS; Thalamus, Th; Amygdala, A; Reticular formation, RF; temporo-parietal junction, TPJ; Temporal lobe, TL; Desolation prefrontal, DP; Orbitofrontal cortex, OFC; Thyroid, Thy.

5.6.11 Love and Pain

Lateral nucleus, LN; Medial ventral posterior nucleus, MVPN; Cingulate cortex, CC; Spindle cells, SC; Olfactory lobe, OL; the limbic system, LS; Superior temporal cortex, STC; orbitofrontal cortex, OFC; Amygdala, A; Primary visual cortex, PVC; Fusiform gyrus, FFG; ventral tegmental area, VTA; nucleus accumbens, NA; Prefrontal cortex, PFC.

5.6.12 Learning Dreaming Defragmentation

Nucleus Accumbens, NA; Thalamus, Th; Caudate nucleus, CN; Reticular formation, RF; Central executive. CE; Prefrontal cortex, PrC; Putamen, P; Parietal cortex, PaC; Hippocampus, H; Broca's area, BA; Visual scratchpad, VSP; Mamillary body, MB; Amygdala, A; Parietal lobe, PL; ventrolateral preoptic nucleus, VLPO; Parafacial zone, PF; Nucleus accumbens core, NAC; Lateral hypothalamic MCH neurons, LHMN.

5.6.13 Fusion of Elementary Sensor into a Single Time Crystal

Optic cranial nerve, II; trochlear nerve (IV); abducens nerve (VI); oculomotor nerve (III); Optic Chiasm, OC; Thalamus, Th; Optic Chiasm, OC; Optical nerve, ON; Retinal ganglion cell axon, RGCA; Lateral geniculate nucleus, LGN; Optical radiation, OR; Cortex domain, CD; Ventral posterior nucleus, VPN; Dorsal column nuclei, DCN; Ventral horn, VH; Dorsal root ganglion, DRG; Touch, Pressure, Vibration, heat/cold, pain, proprioception (muscle), S1–S6; Receptor cell nerve fiber, RCNF; Medial orbitofrontal, MOF; Tract, T; Amygdala, A; Lateral orbitofrontal, LOF; Hippocampus, H; Midbrain, M; Olfactory nerve, I; Facial nerve (cranial nerve VII), the lingual branch of the glossopharyngeal nerve (cranial nerve IX), and the superior laryngeal branch of the vagus nerve (Cranial nerve X); Nucleus of medullary tract, NMT; Taste area of somatosensory, TAS; Taste area of insula, TAI; Medulla, M; Gustatory Cortex, GC; Thalamus, Th; Epiglottis, E; Superior olivary nucleus, SON; Primary auditory cortex, PAC; Superior olive complex, SOC; Mid brain, MB; Lateral lemniscus, LL; Medulla, M; Inferior Colliculus, IC; Ventral Cochlear nucleus, VCN; Semicircular canals, SCC; Auditory nerve, AN; Medial geniculate body, MGB; Inferior colliculus, IC; Cochlear nucleus, CN; Auditory nerve, AN.

5.6.14 Proprioception

Proprioception cells, PC; Pressure, P; Spinal cord, SC; Conscious, Co; Un-conscious, UC; Temperature, T; Cerebellum, C; Thalamus, Th; Vibration, V; Pons, Po; Pain, Pn; Touch, T; Dorsal root axon, DRA; Ventral Posterior nuclei, VPN; Dorsal root ganglion, DRG; Dorsal column nuclei, DCN.

5.6.15 Registering an Event

The putamen, P; Caudate nucleus, CN; dorsal stratium, DS; Substantia Nigra, SN; Globus pallidus, GP; Claustrum, C; thalamus; Optical nerve, ON; Cochlea, Coch; Auditory nerve, AN; Retinal ganglion cell axon, RGCA; Spinal nerve, SN; Cerebellum, Cr.

5.6.16 Homeostasis, Thermal Equilibrium

Pituitary, P; Gonad, G; Adrenal, A; Thyroid, T; Retina, R; Basal Ganglia, BG; Hypothalamus, H; Pituitary gland, PG; Melatomic level, ML; Suprachiasmatic nucleus, SCN.

5.6.17 How Brain Senses the Direction

Semicircular canals, SCC; Cochlear nucleus, CN; Auditory nerve, AN; Optic cranial nerve, II; trochlear nerve (IV); abducens nerve (VI); oculomotor nerve (III), Optic Chiasm, OC; Retinal ganglion cell axon, RGCA; Temporal lobe, TL; Thalamus, T; Parietal lobe, PL; Frontal lobe, FL; Corpus callosum, CC; Reticular formation, RF; Place cells, PC; Hippocampus, H; Grid cells, GC; Entorhinal cortex, EC; Head direction cell, EDC.

5.6.18 Temporal Synchrony of Entire Skin Cover, the Feeling of Body

Somato Sensory Cortex, SSC; Supplementary motor, SM; Raphe Nucleus, RN; Motor cortex, MC; Spinal loop, SL; Insular Cortex, IC; Descending connection, DC; Analgesic chemical cycles, ACC; Parietal Cortex, PC; Pituitary, P; C- fiber, C-F; Thyroid, Ty; Brain stain, BS; Nerve fiber, NF; Dorsal horn, DH; A-delta fiber, AF; Thalamus, Th; Activate automated response, AAR; Spinal cord, SC; Hippocampus, (6 nuclei), H6N; Shiver, S; Voluntary, V; Cerebral, C; Temporal lobe, TL; Thalamus, T; Parietal lobe, PL; Frontal lobe, FL; Corpus callosum, CC; Reticular formation, RF.

5.6.19 Emotion

Medial ventral posterior nucleus, MVPN; Amygdala, A; the limbic system, LS; GABA secretion, GS; Hippocampus, H; Hypothalamus, H; Mamillary body, MB; Olfactory lobe, OL; Lateral nucleus, LN; Fornix, F; Thalamus, Th; Stria terminalis, ST.

5.6.20 Synthesis of Time

Medulla, MD; Pons, P; Internal muscle, IM; Internal muscle, IM; Vagus nerve, VN; Diaphragm, D; Ventral Respiratory Group, VRG; Pontine Respiratory Center, PRC; Dorsal Respiratory Group, DRG; Sinoatrial node, SN; Cardiac nerve, CN; Cardio Regulator, CR; Hypothalamus, H; Basal Ganglia, BG; Dopamin Path, DP; Prefrontal cortex, anterior, PFCA; Substantia Nigra, SN; Pyramidal decussation, PD; Raphe nuclei, RN; Suprachiasmatic nuclei, SCN; Adrenal gland cortisol, AGC; Pineal gland melatonin, PGM; Ventro-lateral preoptic nucleus, VLPO; Retina ganglion, RG; Tubero mammillary nucleus, TMN.

The most interesting aspect is that here, we are not using any biological organs, then how could the circuitry generate a similar response. The answer is that wide ranges of logical output responses are encoded in the circuit. In some logic, multi-channel circuits do not generate responses like real brain components; however, one could calculate the level of complex responses generated by each circuit. Eight junctions in a sequence could have 2^8 choices by switching 'on' and 'off' if there are eight cables in a bundle, it is 2^{64} choices. Therefore, we built 20 circuits of cognitive experiences and connected them to whole-body neural networks built by rigid cables that could be bent and stay in a particular shape as desired by real human brain-body maps. The new humanoid model is shown in Chap. 6.

5.7 Conclusion: A Journey from Resonance Chain or Polyatomic Time Crystal Based Machines

In summary, all brain elements are time crystal analyzers with the task of developing a pattern of primes and connecting different vibrations of new primes. Principally, memory is primarily a time crystal, but the time crystal associated with decision-making is its origin; One can find the rhythm of a few million years ago as a series that survives for trillions of seconds. Brain Jelly is fit to perform so. Time crystals form three classes in the human brain-body—first, renewal time crystal, second, decision-making time crystal, and third, sensory time crystal.

Circadian rhythms run 24 h in the human body; there is a protein cycle for hours, skin cells change within two weeks, bone cells change in 2–8 weeks, and the time cycle of heart cells is 100 years. Hence, there is a transformation cycle from a particular DNA molecule to the heart. Here we changed that period to rhythm and the whole rhythm to the 3D geometric shape. 3D geometries are connected by phase. So, natural intelligence does not follow the instructions. Geometry is the frame of the number of primes. Adding more symmetry is an option to test the data more.

Here, we have analyzed a pure materials science to analyze- what should be its characteristics if we consider the brain as an awareness of less material architecture. We have seen two main features in the brain; the first feature is that the resonance of all brain components is found in the Hz–GHz frequency range if one checks the Hz–PHz (10^{15} Hz) frequency range. The second feature is that electromagnetic energy at all levels is more effective over the entire structure. The EM field distribution of the overall components of the brain is more reliable. This is the journey we started in 2014 with the resonance chain [110], and now ended with creating polyatomic time crystal model of the whole brain (Chap. 4, [12]). Here we created blind replicas of the brain components, much larger, yet, kept the frequency ratio intact, which ensures that artificial structure and the theoretically studied model have the similar polyatomic time crystal architecture.

We made an extreme effort to understand that the new language GML and PPM could work in real engineered devices. We made 100 times larger replicas and the result was impressive because it appeared to us as if properties remained unchanged.

References

1. Simkó M, Mattsson MO (2019) 5G wireless communication and health effects—A pragmatic review based on available studies regarding 6–100 GHz. Int J Environ Res Public Health 16(18):3406
2. Larik RSA, Mallah GA, Talpur MMA, Suhag AK, Larik FA (2016) Effect of wireless device on human body. J Comput Sci Syst Biol 9:4. https://doi.org/10.4172/jcsb.1000229
3. Politanski P, Bortkiewicz A, Zmyslony M (2016) Effects of radio- and microwaves emitted by wireless communication devices on the functions of the nervous system selected elements. https://doi.org/10.13075/mp.5893.00343

4. Havas M (2013) Radiation from wireless technology affects the blood, the heart, and the autonomic nervous system. Rev Environ Health 28(2–3):75–84. https://doi.org/10.1515/reveh-2013-0004
5. Awada et al (2018) Simulation of the effect of 5G cell phone radiation on human brain. In: IEEE international multidisciplinary conference on engineering technology (ICMET)
6. Indulski et al (1997) The present state of knowledge concerning the effect of electromagnetic fields of 50/60 Hz on the circulatory system and the autonomic nervous system. Med Pr 48(4):441–451
7. Russell CL (2018) 5 G wireless telecommunications expansion: public health and environmental implications. Environ Res 165:484–495
8. Falcioni et al (2018) Report of final results regarding brain and heart tumors in Sprague—Dawley rats exposed from prenatal life until natural death to mobile phone radiofrequency field representative of a 1.8 GHz GSM base station environmental emission. Environ Res 165:496–503
9. Libet B (2004) Mind time: the temporal factor in consciousness. Harvard University Press, Cambridge, Massachusetts
10. Loftus EF (1997) Creating false memories. Sci Am 277:70–75
11. Beggs JM, Timme N (2012) Being critical of criticality in the brain. Front Fractal Physiol
12. Bandyopadhyay A (2020a) Nanobrain: the making of an artificial brain from a time crystal. Taylor & Francis Inc. Imprint CRC Press Inc., Bosa Roca, United States, 9 336. ISBN 10-1439875499. ISBN 13-9781439875490. https://doi.org/10.1201/9780429107771
13. Carroll SM (2017) Why Boltzmann brains are bad. arXiv:1702.00850
14. Boddy KK, Carroll SM (2013) Can the higgs boson save us from the menace of the boltzmann brains?. arXiv:1308.4686
15. Reddy S et al (2018) A brain-like computer made of time crystal: could a metric of prime alone replace a user and alleviate programming forever? Stud Comput Intell 761:1–44
16. Singh P et al (2020) A self-operating time crystal model of the human brain: can we replace entire brain hardware with a 3D fractal architecture of clocks alone? Information 11(5):238
17. Götze W (2008) Dynamics of glass forming liquids. Oxford University, Press
18. Palmer RG (1982) Broken ergodicity. Adv Phys 31(6):669; Bibcode:1982 Ad Phy 31:669P. https://doi.org/10.1080/00018738200101438
19. Wheeler J (1996) At home in the universe. AIP Press
20. Gardner JN (2005) The physical constants as biosignature: an anthropic retrodiction of the selfish biocosm hypothesis. Int J Astrobiol
21. Hawking S, Hertog T (2002) Why does inflation start at the top of the hill? hep-th/0204212
22. Weinberg S (1999) A designer universe? In: New York review of books
23. Bandyopadhyay A, Ghosh S, Fujita D (2020b) Universal Geometric-musical language for big data processing in an assembly of clocking resonators, JP-2017-150171, 8/2/2017: World patent, WO 2019/026983; US Patent App. 16/635,900
24. Schmitz D (2018) Derivation of fiber orientations from oblique views through human brain sections in 3D-polarized light imaging. Front Neuroanat
25. Saxena K et al (2020) Fractal, scale free electromagnetic resonance of a single brain extracted microtubule nanowire, a single tubulin protein and a single neuron. Fractal Fract 4(2):11
26. Gardner J (2002) Assessing the computational potential of the eschaton: testing the selfish biocosm hypothesis. J Br Interplanet Soc 55(7/8):285–288
27. David K, Alexandre P (2004) The Bayesian brain: the role of uncertainty in neural coding and computation. TRENDS Neurosci 27(12)
28. Doya K et al (2007) Bayesian brain: probabilistic approaches to neural coding, 1edn. The MIT Press
29. Friston K (2010) The free energy principle: a unified brain theory? Nat Rev Neurosci 11. https://doi.org/10.1038/nrn2787

30. Di Leva A (2016) The fractal geometry of the brain. Springer Science+Business Media, New York, Springer, New York, NY. 978-1-4939-3993-0. https://doi.org/10.1007/978-1-4939-3995-4
31. Izhikevich EM, Desai NS, Walcott EC, Hoppensteadt FC (2003) Bursts as a unit of neural information: selective communication via resonance. Trends Neurosci 26:161–167
32. Hurdal MK, Stephenson K (2009) Discrete conformal methods for cortical brain flattening. Neuroimage 45(1):S86–S98
33. Nicolelis M, Cicurel R (2015) The relativistic brain: how it works and why it cannot be simulated by a Turing machine, 1.1 edn. Kios Press, Amazon Asia-Pacific Holdings Private Limited
34. Poznanski RR (2015) The relativistic brain by Ronald Cicurel and Miguel L. Nicolelis. J Integr Neurosc 14(03):431–435
35. Lanzalaco F, Zia W (2009) Dipole Neurology an electromagnetic multipole solution to brain structure, function and abnormality. Wellcome trust, Hinxton, Cambridge, UK
36. Nobili R (1985) Schrödinger wave holography in brain cortex. Phys Rev A 32:3618
37. Beck F, Eccles J (1992) Quantum aspects of brain activity and the role of consciousness. Proc Natl Acad Sci USA 89:11357–11361
38. Hameroff SR, Penrose R (1996) Conscious events as orchestrated spacetime selections. J Conscious Stud 3(1):36–53
39. Fröhlich H (1968) Long range coherence and energy storage in biological systems. Int J Quantum Chem 2:641–649
40. Bartol TM Jr, et al (2015) Nanoconnectomic upper bound on the variability of synaptic plasticity. eLife 4:e10778
41. Miller GA (1956) The magical number seven, plus or minus two: some limits on our capacity for processing information. Psychol Rev 63:81–97
42. Singh P et al (2021a) A space-time-topology-prime, stTS metric for a self-operating mathematical universe uses dodecanion geometric algebra of 2–20 D complex vectors. In: Ray K, Roy KC, Toshniwal SK, Sharma H, Bandyopadhyay A (eds) Proceedings of international conference on data science and applications. Lecture notes in networks and systems, vol 148. Springer, Singapore
43. Singh P et al (2021b) Quaternion, octonion to dodecanion manifold: stereographic projections from infinity lead to a self-operating mathematical universe. In: Singh P, Gupta RK, Ray K, Bandyopadhyay A (eds) Proceedings of international conference on trends in computational and cognitive engineering. Advances in intelligent systems and computing, vol 1169. Springer, Singapore
44. Nikolić D (2017) Why deep neural nets cannot ever match biological intelligence and what to do about it? Int J Autom Comput 14:532. https://doi.org/10.1007/s11633-017-1093-8
45. Bandyopadhyay A, Ghosh S, Fujita D (2020c) Human brain like intelligent decision-making machine; JP-2017-150173; 8/2/2017; World patent WO 2019/026984; US Patent App. 16/635,892
46. Cao TY, Schweber SS (1993) The conceptual foundation and the philosophical aspect of the renormalization theory. Synthese 93:33–108
47. Feynman RP (1949) Space-time approach to quantum electronic. Phys Rev 76:769
48. Singh P, Ray K, Fujita D, Bandyopadhyay A (2019) Complete dielectric resonator model of human brain from MRI data: a journey from connectome neural branching to single protein. In: Ray K, Sharan S, Rawat S, Jain S, Srivastava S, Bandyopadhyay A (eds) Engineering vibration, communication and information processing, 1st edn, vol 478. Springer, Singapore, India, pp 717–733
49. Eccles JC, Ito M, Szentágothai J (1967) The cerebellum as a neuronal machine. Springer, Berlin

50. Hurdal MK et al (1999) Quasi-conformally flat mapping the human cerebellum. In: Taylor C, Colchester A (eds) Medical image computing and computer-assisted intervention—MICCAI'99 of lecture notes in computer science, vol 1679. Springer, Berlin, pp 279–286
51. Gao Z et al (2018) A cortico-cerebellar loop for motor planning. Nature 563(7729):113–116. https://doi.org/10.1038/s41586-018-0633-x
52. Bliss TV, Collingridge GL (1993) A synaptic model of memory: long-term potentiation in the hippocampus. Nature 361:31–39
53. Buzsáki G, Bragin A, Chrobak JJ, Nádasdy Z, Sík A, Ylinen A (1994a) Oscillatory and intermittent synchrony in the hippocampus: relevance for memory trace formation. In: Buzsàki G, Llinás RR, Singer W, Berthoz A, Christen Y (eds) Temporal coding in the Brain. Springer, Berlin, pp 145–172
54. Leutgeb S, Leutgeb JK, Barnes CA, Moser EI, McNaughton BL, Moser MB (2005) Independent codes for spatial and episodic memory in hippocampal neuronal ensembles. Science 309:619–623
55. Ferbinteanu J, Shapiro ML (2003) Prospective and retrospective memory coding in the hippocampus. Neuron 40:1227–1239
56. Kröger H (1997) Proposal for an experiment to measure the Hausdorff dimension of quantum-mechanical trajectories. Phys Rev A 55(2):951–66. quant-ph/9702013
57. Wallenstein GV, Eichenbaum H, Hasselmo ME (1998) The hippocampus as an associator of discontinuous events. Trends Neurosci 21:317–323
58. Jonas P, Bischofberger J, Fricker D, Miles R (2004) Interneuron diversity series: fast in, fast out—Temporal and spatial signal processing in hippocampal interneurons. Trends Neurosci 27:30–40
59. Buño W Jr, Velluti JC (1977) Relationships of hippocampal theta cycles with bar pressing during self-stimulation. Physiol Behav 19:615–621
60. Stanley SA et al (2016) Bidirectional electromagnetic control of the hypothalamus regulates feeding and metabolism. Nature 531:647–650
61. Schahmann JD, Pandya DN (2006) Fiber pathways of the brain. Oxford University Press, Oxford
62. Toga AW (2012) Mapping of human connectome. Neurosurgery 71(1):1–5
63. Glasser MF (2016) The human connectome project's neuroimaging approach. Nat Neurosci 19(9):1175–1187
64. van Essen DC (2015) The human connectome in health and psychopathology. World Psychiatry 14(2):154–161
65. Ye et al (2015) The intrinsic geometry of the human brain connectome. Brain Inform 2(4):197–210
66. Shi Y, Toga AW (2017) Connectome imaging for mapping brain pathways. Mol Psychiatry 22:1230–1240
67. Basser et al (2000) In vivo fiber tractography using DT-MRI data. https://doi.org/10.1002/1522-2594(200010)44:4<625::AID-MRM17>3.0.CO;2-O
68. Reckfort J (2015) A multiscale approach for the reconstruction of the fiber architecture of the human brain based on 3D-PLI. Front Neuroanat
69. Axer M et al (2011) High-resolution fiber tract reconstruction in the human brain by means of three-dimensional polarized light imaging. Front Neuroinform 5:34
70. Menzel M (2014) Simulation and modeling for reconstruction of nerve fibers in the brain by 3D polarized light imaging
71. Nolden et al (2019) Tracing of nerve fibers through brain regions of fiber crossings in reconstructed 3D-PLI volumes. Bildverarbeitung für die Medizin, pp 62–67
72. Hein KHM (2017) The challenge of mapping the human connectome based on diffusion tractography. Nat Commun 8:1349

73. Reimann et al (2015) An algorithm to predict the connectome of neural microcircuits. Front Comput Neurosci 9:120
74. Bullmore BD, Sporns O (2009) Complex brain networks: graph theoretical analysis of structural function system. Nat Rev Neurosci 10:186–198
75. Leergaard TB, Hilgetag CC, Sporns O (2012) Mapping of connectome: multi-level analysis of brain connectivity. Front Neuroinform l. https://doi.org/10.3389/fninf.2012.00014
76. Mwangi B, Tian TS, Soares JC (2014) A review of feature reduction technique in neuroimaging. Neuroinformatics 12(2):229–244
77. Robert et al (2015) The contribution of geometry to the human connectome. Neuroimage. Article in press
78. Bullmore E, Sporns O (2012) The economy of brain network organization. Nat Rev Neurosci 13(5):336–349
79. Chen et al (2017) Features of spatial and functional segregation and integration of the primate connectome revealed by trade-off between wiring cost and efficiency. https://doi.org/10.1371/journal.pcbi.1005776
80. Zeki S, Shipp S (1988) The functional logic of cortical connections. Nature 335:311–317
81. Bota M, Sporns O, Swanson LW (2015) Architecture of the cerebral cortical association connectome underlying cognition. Proc Natl Acad Sci U S A 112:E2093-2101
82. Sporns O, Tononi G, Kotter R (2005) The human connectome: a structural description of the human brain. PLoS Comput Biol 1:e42
83. Shah et al (2018) Age related changes in topological properties of brain functional network and structural connectivity. Front Neurosci 12:318
84. Budd JML, Kisvarday ZF (2012) Communication and wiring in the cortical connectome. Front Neuroanat 6:42
85. Horton JC, Adams DL (2005) The cortical column: a structure without a function. Philos Trans R Soc Lond B Biol Sci 360(1456):837–862
86. Mountcastle VB (1997) The columnar organization of the neocortex. Brain 120(Pt 4):701–722
87. Haueis P (2016) The life of the cortical column: opening the domain of functional architecture of the cortex (1955–1981). Hist Philos Life Sci 38(3):2
88. Roe AW (2019) Columnar connectome: toward a mathematics of brain function. Netw Neurosci 3(3):779–791
89. Buxhoeveden DP, Casanova MF (2002) The minicolumn hypothesis in neuroscience. Brain 125(Pt 5):935–951
90. Douglas RJ, Martin KA (1991) A functional microcircuit for cat visual cortex. J Physiol 440:735–769
91. Yumiko Y, Jami LM Dantzker, Edward MC (2005) Excitatory cortical neurons form fine-scale functional networks. Nature 433:868–873
92. Brodmann K (1909) Vergleichende Lokalisationslehre der Grosshirnrinde (in German). Johann Ambrosius Barth, Leipzig
93. Garey LJ (2006) Brodmann's localisation in the cerebral cortex. Springer, New York. ISBN 978-0387-26917-7
94. Gawne TJ, Richmond BJ (1993) How independent are the messages carried by adjacent inferior temporal cortical neurons? J Neurosci 13:2758–2771
95. Noctor S, Martinez-Cerdeno V, Ivic L, Kriegstein A (2004) Cortical neurons arise in symmetric and asymmetric division zones and migrate through specific phases. Nat Rev Neurosci 7:136–144
96. Ren J, Momose-Sato Y, Sato K, Greer JJ (2006) Rhythmic neuronal discharge in the medulla and spinal cord of fetal rats in the absence of synaptic transmission. J Neurophysiol 95(1):527–534
97. Mackinnon SE, Yee A, Ray WZ (2012) Nerve transfers for the restoration of hand function after spinal cord injury. J Neurosurg 117(1):176–185
98. Jalife J, Moe GK (1979) Phasic effects of vagal stimulation on pacemaker activity ofthe isolated sinus node of the young cat. Cire Res 45:595–607

99. Raslau FD et al (2015) Memory part 3: the role of the fornix and clinical cases. Am J Neuroradiol 36(9):1604–1608. https://doi.org/10.3174/ajnr.A4371
100. McClelland JL, McNaughton BL, O'Reilly RC (1995) Why there are complementary learning systems in the hippocampus and neocortex: insights from the successes and failures of connectionist models of learning and memory? Psychol Rev 102:419–457
101. Miller R (1989) Cortico-hippocampal interplay: self-organizing phase-locked loops for indexing memories. Psychobiology 17:115–128
102. Sherman SM, Guillery RW (2002) The role of the thalamus in the flow of information to the cortex. Philos Trans R Soc Lond B Biol Sci 357:1695–1708
103. Steriade M, Deschenes M (1984) The thalamus as a neuronal oscillator. Brain Res Rev 8:1–63
104. Penttonen M, Buzsáki G (2003) Natural logarithmic relationship between brain oscillators. Thalamus Relat Syst. 48:1–8
105. Reinagel P, Reid C (2000) Temporal coding of visual information in the thalamus. J Neurosci 20:5392–5400
106. Morison RS, Basset DL (1945) Electrical activity of the thalamus and basal ganglia in decorticate cats. J Neurophysiol 8:309–314
107. Hawkins J, Ahmad S, Cui Y (2017) Why does the neocortex have layers and columns, a theory of learning the 3D structure of the world. https://doi.org/10.1101/162263
108. Tuthill JC, Azin E (2018) Proprioception. Primer 28(5):PR194–PR203
109. Arshavsky Y, Berkinblit MB, Kovalev SA, Chailakhyan M (1964) Periodic transformation of rhythm in a nerve fiber with gradually changing properties. Biofizika 9:365–371
110. Ghosh S et al (2014) Design and operation of a brain like computer: a new class of frequency-fractal computing using wireless communication in a supramolecular organic, inorganic systems. Information 5:28–99
111. Sahu S et al (2013) Multi-level memory-switching properties of a single brain microtubule. Appl Phys Lett 102:123701
112. Chudacek Z (1989) Changes in the thomographic picture of hands of healthy persons after vibrations of 125 Hz. Cech Radio 43(2):98–101
113. Jebur AA, Abed AY (2013) Effect of vibrations transmitted through hand on human body and blood pressure. Int J Adv Eng Sci Technol 3:47–56
114. Lythgo N, Eser P, Groot P, Galea M (2008) Whole-body vibration dosage alters leg blood flow. Clin Physiol Funct Imaging 29(1):53–59
115. Wang L, Zhao M, Ma J, Tian S, Xiang P, Yao W, Fan Y (2014) Med Eng Phys 36(11):1443–1448
116. Osawa Y, Oguma Y (2001) Effects of whole body vibration on resistance training for untrained adults. J Sports Sci Med 10(2):328–337
117. Cochrane DJ, Stannard SR, Firth EC, Rittweger J (2010) Comparing muscle temperature during static and dynamic squatting with and without whole-body vibration. Clin Physiol Funct Imaging 30(4):223–229
118. Lohman EB, Petrofsky JS, Maloney-Hinds C, Betts-Schwab H, Thorpe D (2007) The effect of whole body vibration on lower extremity skin blood flow in normal subjects. Med Sci Monit Int Med J Exp Clin Res 13:CR71–C76
119. Button C, Anderson N, Bradford C, Cotter JD, Ainslie PN (2007) The effect of multidirectional mechanical vibration on peripheral circulation of humans. Clin Physiol Funct Imaging 27(4):211–216
120. Nahriniak V et al (2019) Studying changes of the effective radius in blood vessels after exposure of lower extremities to periodical mechanical vibrations. J Biomed Phys Eng 9(6):673–678
121. Salvendy G (2012) Handbook of human factors and ergonomics, 4th edn. Wiley, New Jersey
122. Dykes RW (1983) Parallel processing of somatosensory information: a theory. Brain Res Rev 6:47–115; Martin JH (2003) Neuroanatomy: text and atlas. McGraw-Hill, New York
123. Braak H, Braak E, Yilmazer D, Bohl J (1996) Functional anatomy of human hippocampal formation and related structures. J Child Neurol 11:265–275

124. Smith EE, Jonides J (1999) Storage and executive processes in the frontal lobes. Science 283:1657–1661
125. Tyng CM, Amin HU, Saad MNM, Malik AS (2017) The influences of emotion on learning and memory. Front Psychol 8:1454

Chapter 6
Can We Ever Make a Humanoid Bot that Runs by Itself Without Any Software?

6.1 Introduction: Existing Artificial Intelligence is a Slave of the Programmer

6.1.1 Fitting Input and Output is Conceptually Orthogonal to What the Brain Does

We live in a world where everyone rightly suggests that the brain is not a machine, artificial intelligence is no good, we need natural intelligence, we have to learn from the neuron and other biomaterials, and finally, we have to invent the brain's functional mystery to revolutionize brain science. However, after saying all these great criticalities, researchers demonstrate only a better fitting method. This is unfortunate that fixing the output when we train a path ensures that we never create anything new. Most importantly, there are multiple hypocrisies in artificial intelligence research that researchers do unknowingly. We have thoroughly cross-checked the programs before concluding all decisions noted below. No paper is cited because we do not want to target any specific work.

1. Researchers encode multiple exciting choices and let them play with random conditions, confusing readers, suggesting that the automaton selects intelligent decisions.
2. Automatons only optimize input and output. It does not know the significance of input and output. The computer has remained a user's instruction implementation device. All shocking new decisions made by AI and deep learning protocols are first instructed and then forgotten by the user himself.
3. Significance less relation between input and output is given a hypothetical meaning, and the abstract, localized, blindly derived relation is correlated with physical parameters to fake an explainable AI
4. Self-learning means following user intuitively designed paths to find rules to correlate datasets. Relating events is what the brain does directly, while in

P. Singh et al., *Biological Antenna to the Humanoid Bot*, Studies in Rhythm Engineering,
https://doi.org/10.1007/978-981-16-9677-0_6

current AI, a user is never bothered about intricate methods to define an event in an unknown dataset.

5. One of the most frustrating things about AI is a hierarchy of foolishness and algorithmic Alzheimer's disease in a programmer's mind. An automaton creator collects possible outcomes from his own life experiences, imagines why that might have happened, then creates a unit of intelligent decision. Researchers forget what is inside one unit when they integrate several such units and create a hierarchical structure. Then, when it encounters a new prediction as output, it calls unknown output magic. This is what we call algorithmic Alzheimer's. More is the number of layers; more shocking is the magical output of an automaton. We call it a hierarchy of foolishness of us, the programmers, and the users.
6. Emulation, replication, following instruction are not signs of intelligence. Decisions that are made, perceived by programmers, are put together to make look-alike decisions. No one made that decision by analyzing, but in reality, a decision is selected from choices to fake artificial variability and confuse users with fake human-like errors.
7. Complexity does not ensure intelligence. A hierarchical network of conditions allows detailed elaborative selections of learning protocols. The complexity of conditional response confuses the user as if it is human-like intelligence, faking intelligence is the present culture.
8. Artificial intelligence protocols believe that speed and power can deliver everything. Increase the number of lines of codes or instructions to the ultimate limit to achieve human-level intelligence and even consciousness. The entire generation of AI machine builders has a firm belief that if not, millions of engineers could intuitively create all possible outcomes and build a database. A faster and more resourceful machine could reach a decision faster than any human, thus, outperforming humans. Outperforming humans with speed and power is not making an artificial intelligence or artificial brain.
9. No researcher is interested in how does a system structures information, encodes information. It is considered that irrespective of systems and nature of properties, information is bits. The assumption about the unified structure of the universe is a mere imagination.
10. Another assumption is that all events in the universe are linearly connected; the concept of parallelism is misleading. It is a sum of two or more linear series of events. In reality, we consider only two systems interacting at a time. All complex world events are considered as the sum of two coupled events.

6.1.2 If the Brain is a Computer, then It Instantly Writes Code, Fits It, Trains It, Run It

The human brain always responds to new situations and events. Simply it means, the brain does not intuitively generate outputs and fits them. What the brain does is exactly the opposite of what computers do. In a computer, the input and the output are known, new paths connecting them are found, while in the brain, the evolving path is known, which we call phase prime metric, and then whenever the brain takes input, one or multiple outputs are generated.

6.1.3 Designing a Machine and then Using It to Model a Life Form is a Crime

6.1.3.1 Learn Four Fundamentals First About the Brain and the Universe

If we observe the history of machines, all the revolutions started by the first and second world war had only one objective, creating machines as human slaves which followed instructions blindly. When the world wars were over, we did not throw the concepts and revisited them being kind to the brain and consciousness. Rather we used information "bits" and "qubits," Turing tapes, the same good old machines to build replicas of brain functions. When four fundamentals, the structure of information, the information integration principles, the unit machine or device that makes the decision, and the protocol to run a self-operating machine that does not require human instruction will be learned from the brain or universe, we would start understanding the brain and the universe as it is. We proposed in 2014 that information structure is a "resonance chain," or geometric structure made of frequencies, which is very different from bits and qubits. Information integrates by following a set of primes regulated by density of primes, i.e., five 3D structures of the ever-evolving density of states. The unit device is Hinductor, a new kind of fourth circuit element [1], a helical phase structure, which grows within and above to build three layers of nested spirals [2, 3].

6.1.3.2 Defining Consciousness: Fundamental Properties that Convert the Machine into Conscious Life

Finally, the properties that define consciousness are (1) seeing the whole in a part of it, i.e., define self; (2) the whole coexist in multiple forms at a time; i.e., out of body entity or mind; (3) creating the whole in many parts within continuously, i.e., regeneration or redefine self (4) converting the whole into a point, line and then to a sphere that is a point, i.e., repeating a journey from point to a point or whole to a whole, i.e., the whole is a point or feeling of completeness (5) creating infinity as a virtual boundary of an ever divergent whole and feedback from infinity or whole to all parts, transformation of whole or incompleteness (6) creating singularity or undefined-ness by putting an arbitrary number of wholes into a point, i.e., empty or feeling of nothingness; (7) information structure of the whole is its physical structure, single entity prime has become everything; (8) Endless journey of the whole within and above always finds a whole; nothing begins, nothing ends.

In a sentence, an entity that does not contain where, what, why, and when is simultaneously the whole and the point, complete and incomplete, finite and endless, within and above, everything and nothing, and it does not begin or end is conscious.

6.1.4 Brain Reduces Computing Element to Be Smart, Slows Processing Speed to Be Wiser

In order to make decisions, the brain does not collect large database and then slowly reduce them to a few choices. First, increasing the number of choices and then reducing it contradicts the brain's decision-making culture. The brain sees a situation or event and then recreates it using its elementary events, pre-determined integration rules. All that brain has to do is use letters of its language and grammar rules, recreate events it receives from the external world, no need for it to compute.

6.1.5 Synesthetic Technology Holds the True Mystery of a Human Brain

We could get an idea about the information structure of the human brain by looking at the sensory cross-connections. Visual information triggers smell, taste, and effects of sounds. It is not a simple cross-connection between circuits. Considering such connections as leaking of information has been the greatest problem in understanding the human brain. We proposed geometric musical language (GML) that we learned from scale-free symmetry in the information structure of a neuron, its component filaments, and proteins that make filaments [4, 5]. When all sensory systems use a similar information structure where geometric shapes are made of clocks, geometric shapes created in one sensor could trigger a similar shape encoded across the brain [6].

6.2 Every Ten years, We Say that We Are Going to Build a Human Brain

Here we suggest ten protocols that we followed to build an artificial brain.

1. An automaton should reject the philosophy of converting any information structure into a linear form. Instead, we should follow the philosophy that information structure is made of only one variable that generates nested spirals of

different geometries to create everything. Everything means events, properties, invariants, etc.

2. An automaton should invent invariants or conserved laws from the input without training to learn an object, entity, or event. It should not require an output to train.
3. An automaton should decide which events create variables. For example, periodic events and a hierarchical network of periodic events could be a good start. So, we need a new language and sensors that do not read bits but recurring events.
4. An archive of primes. An automaton should have a super symmetry archive wherefrom all possible symmetries and their correlations could be derived. This archive should be fundamental, not to be influenced by humans. An infinite resource of prime numbers is an example of an archive where the density of primes could build patterns related to all possible symmetries.
5. An automaton should link all decisions into an integrated architecture. Discrete logical decisions could never hold astronomical pathways of decisions. Only a 3D geometric architecture could do that. There are many advantages of writing information in terms of geometric structure. Not just many pathways are naturally integrated, but they could be programmed as vibrations of hardware.
6. An automaton should be cyclic; its invariants should be cyclic, the archive of all symmetries (fifteen pairs of symmetries) that create everything should also be created by the ultimate element. Cyclicity of decision structure, decision-making process, all physical and non-physical elements created in the artificial brain should be cyclic.
7. An automaton should be founded where the shape of vibration equates to human perceptions, and those perceptions should equate to prime numbers. We may think about the necessity of correlating entities of different dimensions to a dimensionless entity like points and an entity beyond points, like integers. Correlating entities of all dimensions, all properties of nature equate with each other to be converted to another. They would all appear written in a singular language. It would enable creating an architecture of loops that integrate all entities. The advantage of the architecture of loops is that it runs by itself.
8. An automaton should have a purpose that could never be met. The artificial brain should be designed as a divergent system. We have a proposition to use five 3D patterns of the density of primes that extends to infinity, a never-ending ever-evolving pattern. If we design a system that follows these ever-evolving patterns, that is not sufficient. The ever-evolving pattern has an additional property to generate a self-similar ever-evolving pattern. Thus, the automaton has a purpose; whenever the system point starts diverging from the circular path, a new system point emerges. Together, a never-ending yet virtual static system is the purpose of the automaton.
9. An automaton should ensure the coexistence of multiple alternate structures. Throughout the information structure of the automaton (in reality, information = structure), similar conditions generate different information structures.

Automaton would create an ever-increasing database of centers or locations with alternate possibilities called active centers. The inherent fundamental property of the automaton is to close a loop or pattern connecting events, couple different active centers into a group of groups. Group of groups continuously exchanges active centers.

10. Automation should create major features of the whole pattern into its small part. The automaton should have the property to rearrange multiple local ever-evolving patterns to create a virtually closed system or a circle. We call it feedback from infinity. Feedback from infinity does not mean that we send a signal to infinity and then return to the local part. Creating a miniature replica of the whole pattern starts when all points of diverging patterns start closing in. During the boundary's closure, the boundary pattern's effect is reflected in the local domains. The boundary closure adjustment at all local points should follow a finite number of rules. Local part acquires the ability to regenerate a replica of an automaton. This ability is the foundation of a conscious automaton. It is a network of automatons inside an automaton.

6.2.1 Two Fundamental Philosophical Approaches to Make a Human Brain

For many years, two revolutions have been running in parallel regarding brain research. First-building human brain organoids [7, 8], soft brain tissues using 3D assembly [9], and making lookalike the brain components as cortical folds using elastomeric gel [10]. Second, an accurate mapping of brain components to build a functional brain model that resembles a living human brain in cognitive response [11, 12]. Recently, an artificial brain was built by converting human adult skin cells into a pluripotent cell, an immature stem cell; after that, it (artificial brain) converted into a cerebral organoid [8]. Artificial brain synthesis is problematic. For example, an artificial brain built from skin does not have a vascular system, so the brain does not run, but it is a lookalike. For example, nearly 600 genes causing autism need an artificial brain to test complex brain-body interactions of 600 entities. No supercomputer could simulate all possible chain of events.

The synthesis of a pure biological brain from cells seems very difficult at this moment. The artificial lookalike's brains are either at a single cellular scale or mimic the global appearance. There should be an effort that replicates from a single cellular scale to the largest, i.e., the whole brain. What should be the purpose of growth from a single cell to the ultimate brain structure?

6.2.2 Comparative Review of Different Brain-Building Projects and Start-Up Innovations

Most brain models create a black box to connect the neuron firing to the cognitive reactions. Motivated by Feynmann's geometric [13] and Wheeler's geometrodynamics [14] verbal communication of the universe, we go beyond geometric reasoning to a geometric verbal communication. To verify if the human brain follows a geometric language, we do not look at the pulse stream but the shifting of the resonance frequency peaks and their phase relation in wide ranges of time scales. For that purpose, we have considered every cavity in the brain-body structure as cavity resonators and the solid geometric shapes as dielectric resonators. Folds in the cortex form a network of cavities, here, the cavity resonators are nested within and above. Inside each cavity, the nerve fibers and tracts extend as the connectome model. Here each fiber bundle is a dielectric resonator.

Similarly, the hexagonal close packing of cortical columns in 47 functional regions of cortex membrane has 47 dielectric resonators. Each cortical column is a hollow cylinder, so they are cavities. Inside the cortical column, 29 different classes of neurons arrange as 29 types of dielectric resonators. A neuron membrane forms a cavity; all the filaments inside the cavity form antennas like dielectric resonators. Each filament is a cylinder hence a cavity resonator. The cavity surface is made of protein dimers; each dumbbell shape dimer is a dielectric resonator. The protein dimer is also a cavity resonator inside its secondary structures exist; each secondary structure is a dielectric resonator. The alpha helices are hollow cylinders hence the cavity resonator. However, they are also helical dielectric resonators. In this way, the alternate combinations of dielectric and cavity resonators form the entire brain. All components are connected via a chain of resonance frequencies spanned over a wide range of time. This brain model does not reject any component as useless or emphasize a particular component as a supreme element for information processing. Our objective is to find the geometric correlation between resonances peaks at all nested layers, follow any rule that leads to the self-operation of the system.

Most of Brodmann's 47 functional regions of the brain cortex that deliver all decision-making processes are made of cortical columns arranged in a hexagonal close packing. Using semiconducting and metallic wires, we built a detailed replica of the entire neural network of a brain-body system at a reduced 1:1 million fibers ratio. We built each important brain component as a part of the humanoid bot and separately built 100 times larger structures than the biological structure to play the organ geometry more precisely. The electromagnetic resonant behaviors for each isolated component response were compared with the embedded components in the humanoid bot. Structure files of all brain components were purchased from online sources or freely available from those who mapped the real human brain components.

6.3 Geometric Concept

The geometry of neurons can change their spiking rate, and we cannot simulate the structure of a featureless neuron [15]. It is possible to simplify the shape of a neuron based on its electrical properties [16]. The finest debate about the spiking nature was whether neurons use [17] time depended on plasticity (STDP) using a spike rate code that allows for strong or weak synaptic connections to form, which depends on the spike rate of the neuron [18, 19]. Neurons trigger all conscious features, and their signals can be read by supercomputers, including trillions of operations per sec, as well as thousands of processors. Such hardware would artificially connect neurons in a computer simulation with millions of synaptic junctions in the 3D space [20]. Power consumption is the main problem of this device. It needs 10^5 times more power than the actual human brain. The mimicked geometry of brain cells or biological components is implemented in SpiNNaker neuromorphic hardware to run in real-time using both neuron spiking and synaptic wiring (STDP) [12]. The hardware of the machine uses a large number of electronic circuits to mimic the functional human brain. Still, networks such as a synaptic junction require further optimization to increase the real-time simulation capability of large-scale biological components [11]. Simply the brain model is a cavity that links the spiking of the neuron to its perceptional function. Researchers are always interested in learning the role of geometry in processing the human brain's information [21, 22]. Some fundamental quantities, such as difference force, field, mass, charge, and gravitational, can be described in the geometrical methods [14, 23]. The geometrical language of the human brain is an advanced topic of research. Our humanoid bot is based on cavity and dielectric resonators, following the language of geometrics hidden inside the human brain.

Many prototypes of the programming-based humanoid bot have been created, and many of them are available in the market. Still, there was no single effort to create the particular type of brain-body network of the human. Here, we have built every single component of the human brain by 3D printing using different types of material, and cavity-based components are filled with suitable organic material. Synthetic jelly was conceived to test certain features of consciousness. In the same way, we understand, the other human-like bodies are conscious and can talk and interact with the bot on a random subject. The major target of this chapter is to explore the possibility of creating a human brain that is non-chemical, non-algorithm besides it, and it would be capable of processing unknown and unpredictable events continuously.

6.4 Reverse Engineering of the Brain Components

6.4.1 Essential Artificial Components

6.4.1.1 Basic Principles Used for Designing and Constructing the Brain

We are very clear about the fact that we cannot make an artificial brain. Our objective to create an artificial brain of the humanoid bot subject (HBS) is to verify the supremacy of the polyatomic time crystal model of the whole brain.

1. Instead of creating artificial neuron, we replicated the nerve fiber.
2. Junctions were multiple wireless antennas facing each other.
3. Various dielectrics used in capacitors, like, papers, ceramics, amber, were used here in a suitable shape of a cavity and pure dielectric resonator so that wireless signals passing through, develops a desired phase.
4. We built a software tool so that modulated wave form and the time crystal could be interchanged and we verify if polyatomic time crystal is processing information properly.
5. We created seventeen individual organs (Chap. 5) and whole body as humanoid bot subjects, HBS. Isolated and integrated organs were compared.
6. We consider only 537 classes of clocks, in future it could be increased further; here we took minimum classes of clocks.
7. EEG of cortex was used as marker of HBS thoughts. Seeing, hearing, listening, touching, and tasting, the bot could generate corresponding signals in cortex region, we check location and activity pattern to understand whether it is consistent with the real human subject's EEG.
8. Apart from EEG, vortex condensation and vortex generation ability of the HBS was checked by sending laser. Monochromatic light sends assembly of optical vortices that looks like time polycrystals. A light ring corresponds to a periodic oscillation on the organic fourth circuit element Hinductor which was used to build a cortical column.

6.4.1.2 Complete Design of the Main Components of the Brain Components

Seventeen analogs of the seventeen brain components were selected based on the observation of patterns in their configuration. By an organic synthesis, they all form a filmed composite that produces seventeen PPM, with each element having its PPM [6, 24, 25]. The artificial brain built is a hybrid device consisting of a jelly-based supramolecular structure, small–small 3D nanowires, and many of them non-commercial sensors. Only the sensor receives power from the Arduino device that sends the time crystals through the avatar's body to the brain. We have confirmed this by placing an EEG machine on the bot's brain [6].

6.4.2 Artificial Protein and Protein Complexes like Filaments

A protein is made of secondary structures like alfa-helices and bets sheets, alfa helix, DNA, and all helical assemblies. The fourth circuit element produces the primary time crystal, but it has a finite function. To sense the complex time crystal, many sensor elements must be arranged in such a way that they have different lengths of the helices. The number of elements (helices) participating in a cyclic loop should not be prime but rather integer and follow the rule. For example four elements 6-11-11-6, 3-4-4-3 will form the ring, and three elements 5-4-5, 6-2-6 tringle. The production of such cyclic elements through different transporters at different spatial levels is used to control the dark knots [6]. The characteristics of synthetic materials can be manipulated by editing the number of rings in the helices. The development of life and information transmission takes place through the complexes of many proteins, ribosomes, and microtubules. Previous studies reported several spiral nanowires [6]. One can change the propagation of information by changing its lattice arrangement while avoiding its key dynamic features.

6.4.3 Artificial Neuron

We require many different symmetries within the molecular structure, the activity of which is connected with geometrical relaxation, electronic conductivity, and antenna potential. If we start with a hierarchical helical structure, it is possible that electronic switching will be coded into a helix at the primary geometric level and other levels. We will use those helices to create another elementary structure like the antenna's geometry [6]. Many spiral nanostructures assure ballistic transport, which is the quantum effect [26]. We use the specific arrangement of capacitors in creating spring. Capacitors store charge and produce magnetic flux. Under certain conditions, such features transform the material into a self-radiating antenna. This key point is an important feature of the structure of neurons.

6.4.4 Brain Cavities with Folds Are Essential

The computers are allowed to build new neurons as required, and in many cases, conditional commands change their structure. If a group of neurons needs to copy the peripheral time crystal, then a group of neurons is collected. The brain simulator is filled with sound, image, and other external sensory signals; it also arranges to fire patterns and changes the position of neurons [6]. The software selects the incoming picture as a rhythm. If the rhythm changes, it edits the time crystal as the circuit is modified to fold over the brain's surface, and the fold is accounted for in the software.

6.4.5 Artificial Nucleus

The nucleus is the central part where numbers of symmetries interfere. This means we will encode PPM for prime numbers 2, 5, 7, 11, etc., and then the synthetic jelly will change the structure. However, it is also possible to convert the same synthetic jelly into a group of other prime numbers such as 3, 13, 17, 23, etc., if needed [6]. So as required, the time crystal in the jelly can freeze and remain stable.

6.4.6 Artificial Connectome

Several unpaid 3D printed models of connectome are available online. The cylindrical tube is similar to the thin fibers 3D printing [6, 21]. A nerve fiber map helps connect the synthetic jelly and create a massive network of nanofibers to transmit time crystals from the sensors to the brain.

6.4.7 Artificial Cortex

Brodmann suggested 47 function regions on the upper cortex layer of the human brain. These regions are formed by capillary tubes filled with seven layers of the synthetic jelly and approximately 120,000 capillary tubes arranged in the hexagonal form [6]. Brodmann's 47 functional regions of the brain cortex that deliver all decision-making processes consist of cortical columns arranged in a hexagonal close packing. Connectome nerve fibers select 47 regions for the cognitive response, which is the key to the reverse engineering of a human brain. We have built a replica of a humanoid bot using similar dielectric materials and experimentally verified the theoretically built model. When our brain learns, the assemblies of cortical columns change from an ideal hexagonal close packing to a deformed one. It means the lattice parameters would change as the brain learns. We have experimentally estimated how conscious experience occurs at the humanoid bot's brain by interacting with another human and how the geometry of hexagonal lattice is related to the emergence of consciousness in a human brain. Our entire study is a non-biological and transmission line circuit element based on reverse engineering of the brain components. However, brain components are meticulously built from old reports of brain architectures. Thus, it is a primitive yet first step to test hypotheses on the brain rapidly.

6.4.8 Artificial Cranial Nerves

The cranial nerves are part of the PNS (the peripheral nervous system). It organizes the sensory inputs essential for life sciences. Such sensory inputs go to the midbrain and are identified by a specific type of dielectric material there. C2, C3, C5, and C7 symmetries are found in the spinal cord and cranial nerves [6]. Fractions that use the same symmetry are marked through the same dielectric material.

6.4.9 Artificial Cortical Column

A single cortical column has seven layers. Interestingly, the input signal enters the brain's cortex from the top of the available column and exits at the bottom. We have created 27 types of distinct neurons and arranged them in different types of cortical columns. A group of H (Hinductor) devices forms neurons, self-assembles, and behaves as a cylinder [6, 25].

6.4.10 Artificial Basal Ganglia

In synthetic organic jelly, the basal ganglia have thirteen elements similar to supramolecular structures. A similar structure of this type is produced from the 3D printer.

6.4.11 Nineteen Artificial Brain Stems

Seventeen nuclei are used to make the brain stem. Every nucleus is specific to process a specific prime number.

6.4.12 Artificial Cerebellum

Often the structure of the cerebellum is like a fractal tree chosen as a cavity (see Chap. 5). The cerebellum structure consists of two vertical and horizontal structures, and two supramolecular structures intersect in the middle regions.

6.4.13 Artificial Hypothalamus

The hypothalamus is similar to the thalamus, but a geometric feature is that all structures such as viruses rule wherever there is symmetry, so the desired program can be derived in the supra-molecule.

6.4.14 Artificial Spinal Cord

The spinal cord is a complex network of nerves that are connected to the skin nerve network. Conductor gel fibers are formed near the area where signals generated from thermal, pressure, and non-contact sensors make a mixed effect and form the same shape as the backbone. It is made from 3D printing. Holes are self-assembled into gels to make connections between spinal input fibers and motor networks parallel to each other.

6.4.15 Artificial Limbic System

A high-radiation material is used to form the limbic system. Thus the material reacts like a transmitter and receiver.

6.4.16 Artificial Thalamus

Fundamental geometrics are encoded in the thalamus, and incoming signals are programmed over the entire operational range. The corresponding time crystals are generated in the artificial brain, and the aggregates or integrations of specific geometries are powerfully inserted here [6].

6.4.17 Artificial Blood Vessel

3D printed blood vessels are made using elastic material, and by pumping ultrasound waves, it floats in the brain like a real network.

6.4.18 Artificial Hippocampus

The hippocampus has an encoder and a resource antenna that must always be stable during the process of finding and testing. To find the original information, we have added two buffer antennas for communication inside the hippocampus. In the same way, a hierarchical antenna can modify the time crystal itself on the cortex layer following the law of transformation, although it requires another buffer antenna. The process of time crystal transformation is the dominant feature of artificial brain operation. Such a procedure must be different for both the forebrain and midbrain or for more than two antennas. This process requires two antennas, one to take input from the 'resource antenna and higher brain' and the second to take the output from the buffer antenna. An artificial hippocampus that we made consists of four classes of giant antenna controlled simultaneously. The test of the 'finding process' is followed by better scale rules so that there is no disturbance in the operation of the brain. In short, the artificial brain has two layers of antennas inside each neuron. Then there are four antennas pairs: dual transformation antenna pair, encoder antenna pair, hierarchical antenna pair, and resource antenna pair [6].

6.5 Constructing the Brain Components

Electromagnetic interaction response of bot's brain with nerve fiber bundles is measured in term brain waves: delta, theta, alpha, and beta wave. A schematic and background images of the actual experimental setup are shown in Fig. 6.1. Figure 6.1a shows the brain nerve fibers maps with 47 Broadmann's cortex regions (white circles) to study brain wave's EEG spectrum when signals are pumped wirelessly through 10 function generators (35 kHz–4.4 GHz) channels. A programmed Arduino board controls all channels. Input signals of a fixed bandwidth acquired by 43 helical antennas connected to 5 nerve fiber bundles through the midbrain region. Output responses of brain nerve fibers are measured by an EEG machine placed on thermoplastic, based on a mimicked map of the upper cortex layer. The artificial connectome model has a volume of 1525 cm^3. Both transmitting and receiving helical antennas are aligned in the transverse direction of the connectome model and coupled with a distance of 28.5 cm and 6.4 cm from the ground surface and function generator, respectively. EEG spectrum and electric field distribution are measured during 'on' (Figs. 6.2b and 6.3a, left panel) and 'off' position of all RF channels (Figs. 6.2b and 6.3a, right panel). Variation of the beta wave on the brain cortex surface is measured with time (60 s) to understand various functional parts of the brain (Fig. 6.3b).

Interaction of the electric field with the brain cortex surface is also measured on the humanoid bot's brain. Photographs of the humanoid bot and actual experiment setup are shown in Figs. 6.4 and 6.5. The humanoid bot consists of 20 conscious brain circuits (see Chap. 5) includes the features of all brain components like the hippocampus, cerebellum, midbrain, limbic system, etc., and entire neural body

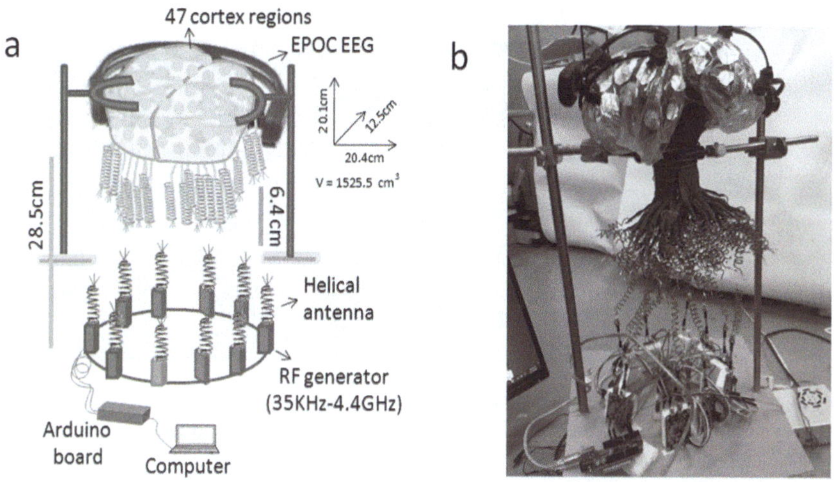

Fig. 6.1 The schematic and experimental setup of the artificial brain model constructed to test the handmade humanoid bot brain is shown in panels (**a**) and (**b**), respectively. An artificial brain model (volume = 1525.5 cm^2) built using thermoplastic contains a mimicked map of the human brain nerve fiber or connectome model. The brain model consists of 47 cortex regions or Brodmann regions, and the whole-brain model is triggered by a channel of ten RF frequency sources (35 kHz–4.4 GHz). EPOC EEG measures output responses or brain waves

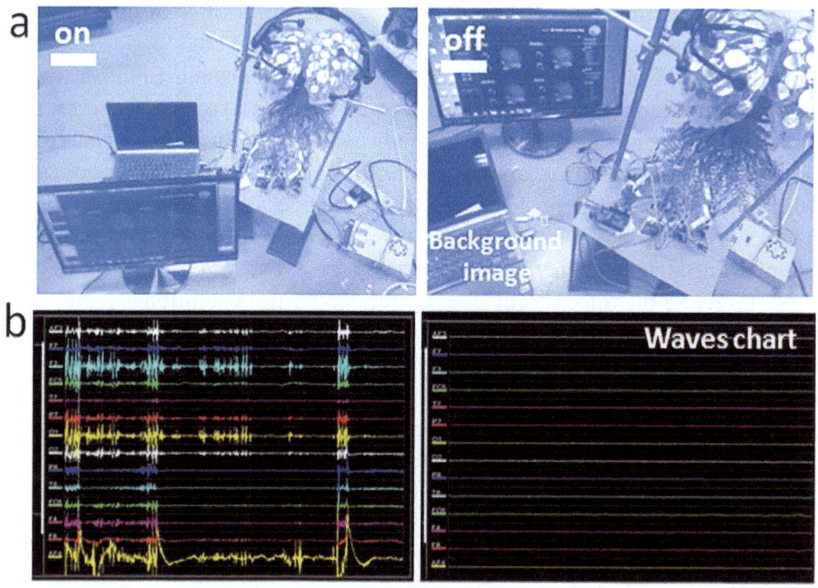

Fig. 6.2 The background image of the experimental setup to measure brain waves by artificial models is shown in panel (**a**) when the RF input sources are switched on and off. At the same time, the corresponding brain wave charts are displayed in panel (**b**)

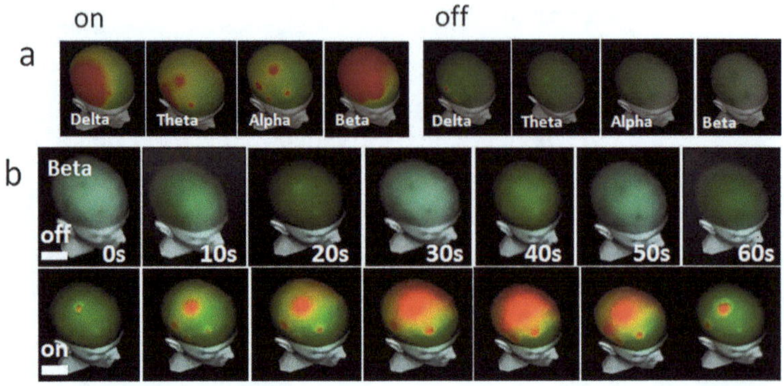

Fig. 6.3 Panel (**a**) shows a chart of brain waves when the power source is turned on and off, and the varying profiles of the beta wave for a period of 0–60 s are shown in panel (**b**). Brain waves such as delta, theta, and alpha dominate the left and right temporal lobe, while beta wave appears over the entire brain region except for the frontal lobe (RF generators-on, panel a, left). All brain waves are silenced during the switched-off of the RF generator (panel a, right). Panel (**b**) shows the time-varying profile (for 0–60 s) of the beta wave at switch off and switch on of RF channels. Beta wave is almost silent for the entire time (0–60 s) domain when RF sources are off, but it starts to appear on the parietal lobe during 20–50 s when RF sources are on

network that includes the 31 pairs of the spinal cord, 12 pairs of cranial nerves, and 12 pairs of thoracic nerves. However, we also added five kinds of human sensors: visual, touch, motion, sound, and position sensors on the bot's body. To study of natural responses of the humanoid bot is observed by placing EEG on the bot's brain and connecting a multichannel analyzer (34 channels) to spinal nerves. Planer spiral antennas are positioned on the bot body to absorb the available signal from the surrounding space and are also responsible for generating the unified information of the spinal cord and brain EEG spectrum. Include all sensors and entire body networks driven by a 5 V power supply from the Arduino board. We recorded the maps of electromagnetic energy distribution on the bot's brain by shifting 16 probes on the surface of the upper cortical layer (thermoplastic) conditionally. All sensors are getting switched 'on' and 'off,' as shown in Fig. 6.5.

The brain exhibits a complex operation. When two or more operations happen simultaneously, then they have some common pathways. When the human brain performs a particular operation, some brain circuits activate, and their final effects appear on the cortex layers, which are detected by EEG. Simultaneously, if additional operation happens, brain circuits activate parallel and change energy distribution patterns or brain waves on the cortex layer. Thus, the first operation changes cause the second operation. For example, if two sensors simultaneously activate, the noisy profile of brain waves appears because they have common pathways in the midbrain region from which it divides into different brain regions.

Their clocking model can represent the fusion of all signals sensed on the skin nerve net and spinal cords (see Chap. 5). When a signal passes through any brain

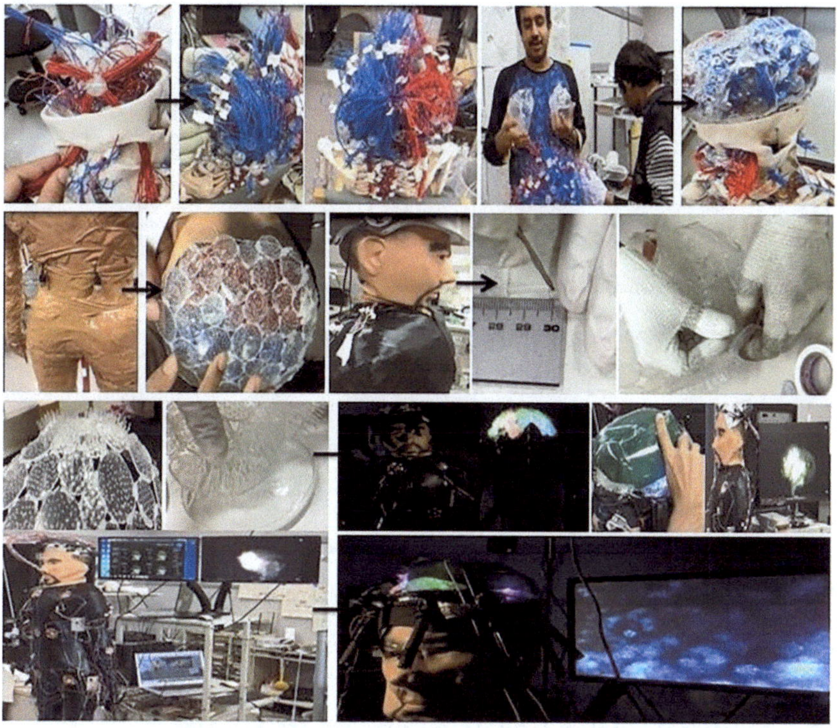

Fig. 6.4 Step by step construction of humanoid bot. Its outputs are detected as EEG and optical/magnetic vortices. The nervous system of the humanoid brain consists of 20 brain conscious circuits (as reported in Chap. 5), 31 pairs of neural spinal nerves, 12 pairs of thoracic nerves, and 12 pairs of cranial nerves with detailed geometric features [6, 25]. The humanoid brain is triggered by an external applied Yagi antenna in the given frequency region (35 MHz–12 GHz). Fluctuating brain waves are generated by triggering the humanoid bot's visual, touch, motion, sound, position, and temperature sensors. Brain waves are detected by placing the Epoch EEG on the bot head

component, it passes through its many sub-components, sub-sub components, sub-sub…sub-components, so signal time during these sub-components are different. The sum of these times is equivalent to the signal period passing through the entire component. So it is one big clock made by fusion of many clocks. A particular example is the creativity and humor circuit built inside the Bot's (see Chap. 5), where the junction is mimicked by a plastic ball connected to several wires. All signals wirelessly communicate through built Yagi antennas at the end of wires. Parallel activation of all or a few clocks in the brain is always mysterious. All clocks are locked based on phase and period. Anyone could easily detect output responses by activating or deactivating a few or more clocks.

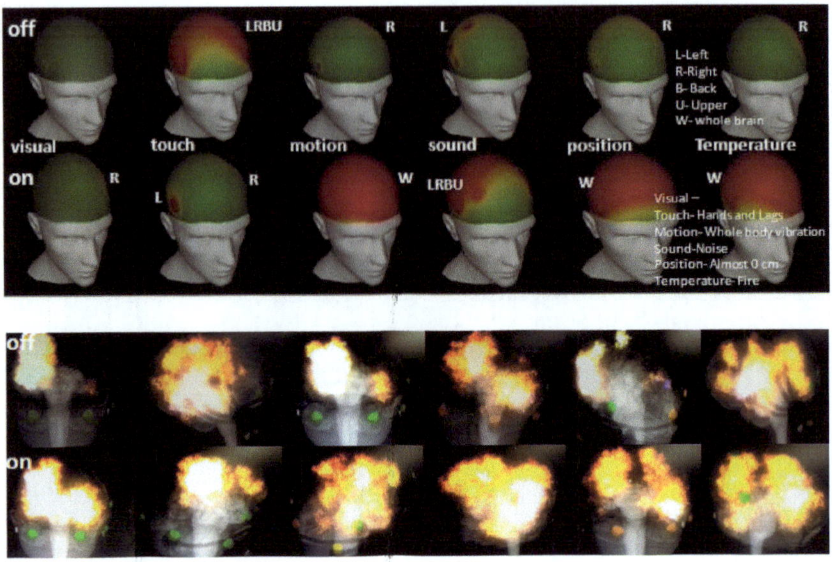

Fig. 6.5 Comparative profiling of beta wave (top panel) and neurons firing pattern (bottom panel) during switch on and switch off of visual, touch, motion, sound, position, and temperature sensors attached on humanoid bot body. Different brain regions are activated for each sensor

6.6 Artificial Cortex Construction and Its Studies

We built an artificial brain model that includes 47 Brodmann cortex regions and the connectome model [27, 28] via two methods. The first route is creating the five major regions of brain fibers as found in the real human brain (Fig. 6.1a) that is replicated meticulously detailed in Figs. 6.4 and 6.5. In Fig. 6.1a, there is no nervous network of the whole body. Instead, we sent the signal wirelessly by an RF network. The simulated signal is background corrected by applying homogeneous bursts of a fixed signal to the connectome network. The real signal means cognitive data; millisecond biological responses were normalized in 35 kHz–4.4 GHz domain similarly as natural signals are sonified. On the top part of the connectome model, the upper cortex layer is created using thermoplastic to place the EEG machine for reading electric field distribution in 47 brain functional cortex regions. The actual experimental device is shown in Fig. 6.1b. An Arduino board controls ten function generators to send the desired signal to an artificial midbrain structure. From the midbrain, 43 helical antennas were extended to acquire input like a biological brain. Figure 6.2a shows how two different computers were used for sending signals and acquiring the EEG as raw data from multiple channels (Fig. 6.2b) and simulated output of connectome in 47 cortex regions.

We know how 47 regions that execute the brain's cognitive functions are distributed [6, 25, 29–31]. Therefore, we could see which part of the brain was getting active in the EEG spectrum, as shown in Figs. 6.3a, b. In general, the experiment is straightforward; by shifting 16 probes, we checked the EEG activities of various functional domains. Here we find that if all channels in RF generators are switched off, there is no EEG response. The geometry of wiring alone could generate the required functional responses as expected. There are two interesting aspects of this experiment. First, we could replace ionic biology that operates in milliseconds time domain and switch to unique hardware that operates in radio and microwave domain and still generate a similar EEG. The effort is primitive but delivers confidence to construct much more biologically accurate and complex brain architecture in the future. The second aspect is that we can solve Maxwell's equations for complex dielectric structure-based hardware designed theoretically. Therefore, experimental verification of theoretical hypotheses for brain function, cognition, and decision-making would be possible for any brain model in the future.

The next step is to confirm that the functional domain responses obtained from the biological connectome replica are not an artifact. Therefore, we went back to the massive brain literature and collected how different cognitive responses of a conscious brain works. The brain component pathways have been documented, and we selected 20 well-documented conscious experiences of a human being found in any classic biology textbooks [6, 29]. We have presented the operational pathways in terms of nested clocks (see Chap. 5), as shown in two cases of Figs. 6.1 and 6.5. There are two reasons for using such a new protocol. First, we could easily build a circuit that would naturally trigger signal processing logically. Developing self-operational hardware where not a single line of code is important because, from the starting point, the most precious feature of a human brain responding to unknown situations should be honored. Second, conscious responses were presented in a linear pathway; however, simultaneous multichannel networking happens during a real brain operation. When we convert linear circuits into a loop, one could use junctions to trigger several interconnected circuits with less complexity. For example, conscious circuits execute the function of a real human brain. Many antennas coming from different directions face each other at the junction.

We have replaced the blind replica of connectome structure from a built model, as shown in Fig. 6.1, with 20 conscious circuits. However, all the circuits were connected to Yagi antennas following the connectome geometry. A total of 120,000 antennas were created, which sent the signal to the cortex layer. Another major change in the model of Fig. 6.1 is that an intricately built neural network, including a real lifelike network of 43 peripheral nervous systems (PNS). All components, spinal cord, midbrain, cerebellum, hippocampus, etc., were built eventually to connect the 20 operational circuits of the upper brain. Here we did not provide the rigorous details of the intricate fabrication process; however, a summarized detail is given in the previous Chap. 5. Five types of human sensors are added to the humanoid bot's body, as shown in Fig. 6.4. We have cross-checked that all sensors send signals naturally when someone speaks in the lab or moves near the humanoid bot. We observed the bot's EEG and spinal cord responses by the multichannel analyzer 24×7 for weeks. The humanoid bot has no software controller; touch, thermal, pressure, auditory, taste, and visual sensors are run by a 5 V power supply from

Arduino, which sends a signal to the neural network of HBS in terms of spike-like pulses. Therefore, the spontaneous change in the laboratory is the only source of the input signal to HBS. Implemented spinal cords integrate the information and EEG output. Such an example is shown in Fig. 6.5. On top of thermoplastic, we made 47 regions filled with 10,000 capillary tubes whose inputs are wirelessly transmitted signals from connectome edges (see Fig. 6.4).

We have compared the ratios of resonance frequencies of the artificial brain with mm^3 volume for two distinct models, one shown in Fig. 6.1 and another in Fig. 6.4. Since we have theoretically studied the connectome structure and experimentally built it, at this point, we look back into the cortical column bundles. We studied how electric and magnetic field distribution change when a similar input is applied to two distinct connectome models (Figs. 6.1 and 6.4). By shifting the localized positions of cortical columns, one could generate electromagnetic energy distribution on the cortical column assembly similar to as presented in Chap. 5. Using the electromagnetic sensor, we could read the point-by-point field distribution of the brain cortex. The learning process in the brain alters according to cortical column position. We need to do it manually in the humanoid bot to verify the experiment. Currently, we are putting helical carbon nanotubes within capillary tubes, which act as a marker of electromagnetic field shifts required for a given pictorial or visual input to the brain.

6.7 Reverse Engineering of Emotion Centers in the Brain: Thalamus, Amygdala, Hypothalamus

Decisions hardly remain logical when the output from the cortex region passes through the emotion control centers in the thalamus, amygdala, and hypothalamus. We have created human brain-like components artificially in the software simulator and reality for physical testing of the device. Twelve thoracic nerves in the chest are not equated with human emotions. However, breathing changes significantly with the emotions of a human being. Hence we have also created a humanoid bot whose thoracic nerves experimentally and three components, thalamus, amygdala, and the hypothalamus, were studied together. By experiment and theory, we have analyzed the mutual effect on key brain circuits for sensory signals, direction sense, fear, threat, anger, pain, and love located in the brain of the humanoid bot.

6.7.1 Thalamus, Amygdala, Hypothalamus: Combined Rhythms of Emotional Stress Under-Regulated Breathing

Initially, the limbic system is defined by the Paul D. MacLean that is the part of the human brain, located at both sides of the thalamus just below the temporal lobe of the cerebellum [32] and above to brain stem, involve with the memory, emotion, behavior, and motivation. Hippocampus and amygdala are two major components of the hippocampus, although some other components like the thalamus, hypothalamus, and basal ganglia are also involved in its action. The limbic system is not an isolated entity; it is the combination of many parts like the fornix (antenna-like shape), the column of the fornix, cingulate gyrus, olfactory bulbs, mammillary body, etc., which have the function to serve several different functions like time perception, consciousness, attention, memory, attentions, etc. Schizophrenia and epilepsy are disorder associated with the limbic system [33].

6.7.1.1 Thalamus

The cerebellum is a small structure located between the midbrain and cerebral cortex and linked by nerve connection. Brain stem is located above it. The thalamus is the relay station of signal processing in the brain, the composition of different types of nuclei, performing the specialized task for various types of information. During brain activities, nearly all signal passes through the thalamus. Wakefulness, sleep, and alertness are the functions of the thalamus.

6.7.1.2 Amygdala

Amygdala looks like an almond shape, made from the clusters of nuclei and located in the temporal lobe of the brain. It exhibits the main role in the decision-making process, memory, and emotional responses. Emotions may be conscious and unconscious features, but we are unaware of them most of the time.

6.7.1.3 Hypothalamus

The hypothalamus located below the thalamus has a large number of the nucleus that perform various functions. Hypothalamus links the endocrine system and nervous system by the pituitary gland.

6.7.2 Loop Pathways of Signals in Thalamus, Amygdala, Hypothalamus, and Cortex Layer

6.7.2.1 Emotion May Be Conscious and Unconscious Feature, and Its Action Pathways

As we well know, we have the emotional center in our brain, but we always keep the primary responses from the heart. The thalamus picks the sensory information (any dangerous event or sound) and sends it to the amygdala to respond quickly. Amygdala accesses the information and sends it to other areas for taking immediate action. Amygdala sends the signal to the hypothalamus, which carries the suitable changes in hormones that make it ready to the body to take the reactions against the emotional stimuli. Such responses change our physical activities like increasing the heart rate and variation in muscle contraction, known as an unconscious emotional pathway.

Suppose the thalamus sends the sensory information to the sensory cortex for recognition. In that case, information consciously enters the hippocampus, encoded in the form of memory, hippocampus confirms or modifies the information. Information is a polyatomic time crystal, a 3D clock assembly, where multiple distinct clocks are used. When only onc typc of clock is uscd, we say it is a time crystal because if the clock runs, we will get many atoms in the temporal domain. However, when multiple atoms or clocks run together, the scenario changes totally. The reason is the relative phase relationship between the clocks.

6.7.3 Signal Pathways in the Individual Thalamus, Amygdala, and Hippocampus

6.7.3.1 Thalamus

Three basic cells—interneurons (GABAergic-GABA as a neurotransmitter), relay cells (glutamateric- glutamate as the neurotransmitter), and reticular cells (interneurons-relay cells are their target), are involved in thalamus processing, and they contain their subtypes. Relay cells receive the input source from the information source and transmit it to the cortex. During this time, the axon passes by the thalamic reticular nuclei and produces the branches which influence reticular cells. The influences of the interneurons are limited at the thalamic site. Interneurons provide the inhibitory connection with the relay cells. Relay cells receive two inputs from the interneurons. Interneurons make more effect on dendrites locations than the reticular nucleus. A study [34] shows the functional circuit between the thalamus and cortex. It describes the connection among those 3 basic cells by two extrinsic inputs to the thalamic circuit, one from a group of cells in the brain stem (brain stem cell group). The second is from the 'glutamatergic and arises from the cortex layer 6'. Second extrinsic input is related to the reverse feedback projection because it innervates the same relay cells which innervate the cortex area. Observing the effect of the cortical input on the relay cells is a little challenging task. Sherman et al. [34] reported that the cortical input directly excites the relay cells and indirectly makes the inhibitory circuit with them and other cells like thalamic reticular cells or interneurons.

6.7.3.2 Amygdala

Amygdala sends and receives the projection from multiple brain regions like the reticular nucleus, hypothalamus, dorsomedial thalamus, facial nerve trigeminal nerve, laterodorsal tegmental nucleus, locus coeruleus, and ventral tegmental area. Amygdala makes the circuity connection with the sensory regions, conditioned response, memory, executive and decision function, emotion, pathology, and dysfunction. McFadyen [35, 36] shows the neural signal processing, white matter, and structural-visual relationship between pulvinar and amygdala. The amygdala plays the primary role in the visual signal processing that can adjust the functions of the cortical column network during the visual stimuli. Under this processing, the cortex has a significant role in visual processing. The pathway passing through the pulvinar and superior colliculus to the amygdala has a prominent role in effective visual stimuli processing [37]. Subcomponents of the amygdala offer coordination regulation during attentional processing [38]. Some studies reported the memory circuity within the amygdala and its interconnection with the brain region on the animal model [39]. In the last decade, several studies have made progress on the specify the brain circuits to emotional functions, many of them based on the classical fear conditions -conditions which are used to produce the environmental stimuli to emotional response circuits where the amygdala is the linkage element [40].

6.7.3.3 Hypothalamus

Hypothalamus is important in controlling the endocrine system, maintaining the body functions, metabolism, and some other conditions involving human behavior. It is more important in the homeostatic of human organisms in blood flow, body temperature, weight, etc. Hypothalamus makes the most complex circuity connections with the brain region by neural and non-neural communication pathways. Hypothalamus links with the midbrain, thalamus, amygdala, hippocampal region, olfactory bulb, retina, and cerebral cortex [41]. By the bidirectional and uni-directional pathways, for example, the hypothalamus and amygdala are interconnected by the ventral amygdalofugal and steria terminalis pathways.

6.7.3.4 Amygdala Connections with Sensory Circuits

Amygdala makes the bidirectional projection with the sensory primary cortices, sensory association cortices, modulatory circuit, polymodal sensory nuclei, memory circuit, decision circuit, and emotion circuit. It sends the information to the autonomic circuit however receives from executive circuits and sensory thalamic nuclei.

6.8 The Construction of an Artificial Spinal Cord and Its Related Studies

6.8.1 Natural Sensory Signal to the Spinal Cord

Our brain is located at the top of the body, which directs and coordinates all activities throughout the body. It happens by spinal cords and nervous system that branch out in network form distributes throughout of body. The spinal cord carries the information from the whole body to the brain and brain to the whole body by dorsal roots (a type of fiber). Cranial nerves serve as the information carrier in the head. The spinal cord combines nerve fiber bundles that is a protuberance of a nerve cell. The spinal cord extends from the brain's base to the bottom part of the spine. The spinal cord has minimum width; the data from different body parts is collected via spinal nerves and transmitted to the brain via the spinal cord. It also serves to transmit the movement coordinate from the cerebellum to our body. Thirty-one pairs of spinal nerves branch out from the spinal cord [6, 29]. Our limbic system plays a significant role in incorporating the visceral and cognitive functions and a connection set-up engaging the amygdala and hippocampus. The connection of the limbic system involving the spinal cord and brainstem is not defined, such connection leads to new ideas regarding the periphery and central nervous system of the body. Pathways regarding the human brain's spinal cord and brain stem are confirmed [42]. How features of brain-body components get the effect by 5G wireless signal is our topic of discussion?

The spinal cord acts as a natural antenna. The spinal cord and electromagnetic wave analysis is carried out in FDTD simulation [43]. Maximum activity of the spinal cord has been seen at the frequency peak, which has a significant magnitude of current and voltage at FM frequency 100 MHz. The vibration occurs in the central nervous system and blood–brain barrier (BBB) at the resonance peak. In the 1979s, the high exposure to EMF caused deaths [44]. Epidemiologic research addressed the role of the electric and magnetic fields in the etiology of cancer at 60 Hz.

The spinal cord is the prime component of the CNS and is attached to the encephalon. The spinal cord integrates and processes the information between brain and body. The spinal cord consists of a sensory nerve and motor nerve. Sensory and motor nerves carry the information toward and away from the spinal cord, respectively. The spinal cord can alone perform the function of the encephalon. It is a signal transmission pathway between cerebrum and body, communication between them is not possible [43]. Spinal cord biomaterial shows the resonance characteristic in the radio frequency range. All the biomaterials are insulators that do not have a highly reflecting outer boundary, so electromagnetic signal reflects multiple times and generates high-quality standing waves). The electrical activity of the spinal cord could be an image by generated magnetic field using sLORETA (spatial filter version of standardized low–resolution brain electromagnetic tomography) [45]. Magnetic stimulation of CNS and PNS has led the therapeutic interest. The neuron mathematical model is helpful to understand the possible effects occur by magnetic stimulation and has clinical applications [46]. Magnetic resonance spectroscopy is used to detect neurologic disease and spinal cord injury, yet this technique methodology has challenging resolution and signal quality [47, 48]. Leg muscle activity could be explained by electromyography, where we make magnetic stimulation of the spinal cord. An electric field is induced in the spinal cord and circumventing region [49].

6.8.2 The Construction of Spinal Cord

6.8.2.1 Theoretical Construction of Spinal Cord: Discussion of Detected Resonance Curve with the Energy Distribution Along with the Structure

We theoretically simulated spinal cord geometry in antenna simulation software by following its actual biological details (Fig. 6.6a). The reflection coefficient is the ratio of the relative amplitude of the incident wave to the reflected wave at the input port. In contrast, the transmission coefficient is transferred amplitude at the second port when a wave of particular amplitude is applied at the first terminal. The resonance spectrum of the spinal cord with two-port is shown in Fig. 6.6c. The spinal cord offers an interesting energy profile. At both resonance frequencies, energy is transmitted from bottom to top port. However, due to port 2, EM energy does not transmit top to bottom; it concentrates on the mid brain region (Fig. 6.6d).

6.8.2.2 Experimental Construction of Spinal Cord: Variation of Signal Intensity at Different Resonance Frequencies at 0°, 90°, 180°, 270° Angles Along With Different Parts of Spinal Cord

We have purchased the plastic mimicked model of the spinal cord and fixed the flexible semiconductor wires in the form of spinal nerves, which could easily bend to take any shape. The thickness of the cable is identical to nerve fiber. We built 31 pairs of the spinal cord at both left and right sides, and the open end of the sensory nerve has a Yagi antenna shape. All spinal nerves are stimulated by a channel of 10 radio frequency sources (35 kHz–4.4 GHz) which are controlled via an Arduino board, and results are detected by a spectrum analyzer (300 kHz–4.4 GHz) (Fig. 6.7a) at the

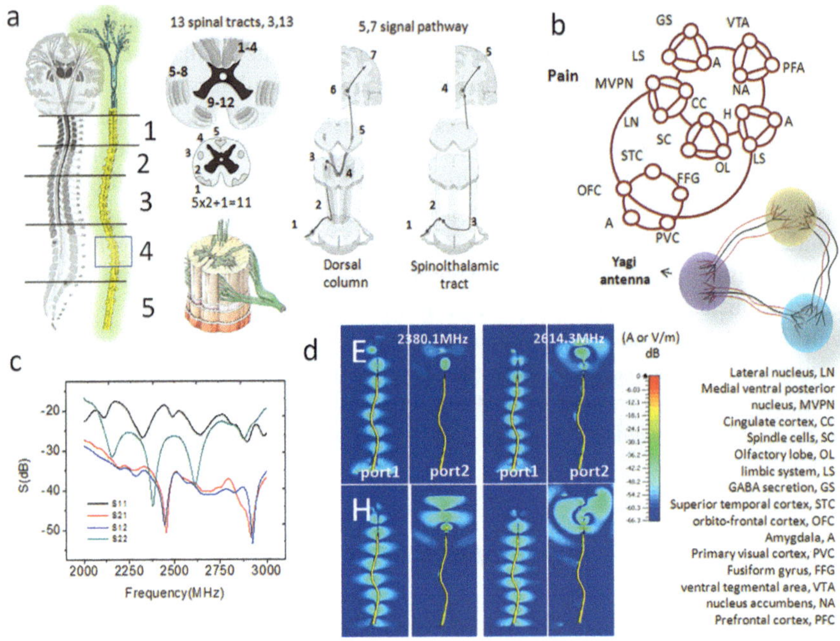

Fig. 6.6 A mimicked model of the spinal cord with five sub spinal regions; cervical (1), thoracic (2), lumber (3), sacral (4), and coccyx nerve's region (5) is shown in panel (**a**). 31 pairs of spinal nerves consist of 5 sub spinal regions. Eleven different regions are found in the spinal cord. 13 spinal tracts pass and collect data from different parts of the human body and send it to the brain through 5 dorsal or 7 spinothalamic signal pathways. The clocking model of the 'pain' conscious circuit is shown in panel (**b**) top, while the bottom part presents a schematic of triangular clock architecture made by wires whose open-ends act as Yagi antenna. The spinal cord's resonance spectrum (S11, S21, S22, and S12) is observed in MHz frequency, 2380.1, and 2614.3 MHz f (panel: **c**). The electric and magnetic field is observed at resonance peaks (panel **d**). **Simulation details***:* Used solver—Maxwell equation solver; selected mode—time domain; boundary condition—open space; waveguide port dimension—10 cm × 10 cm; frequency domain 2–3 GHz range

perpendicular direction to the spinal cord (0°, 90°, 180°, and 270°). Signal intensity and resonance peaks of spinal cords are observed at 4 of its different parts assigned by different colors, as shown in Fig. 6.7b, top and corresponding plots are shown in Fig. 6.7b, bottom. The color coding for both panels is the same. The intensity plots for all 4 regions are almost the same, but frequencies curves vary at various angles for all regions (Fig. 6.7c). Thus, the spinal cord is perfect dielectric and acts as a transmission line with no energy loss.

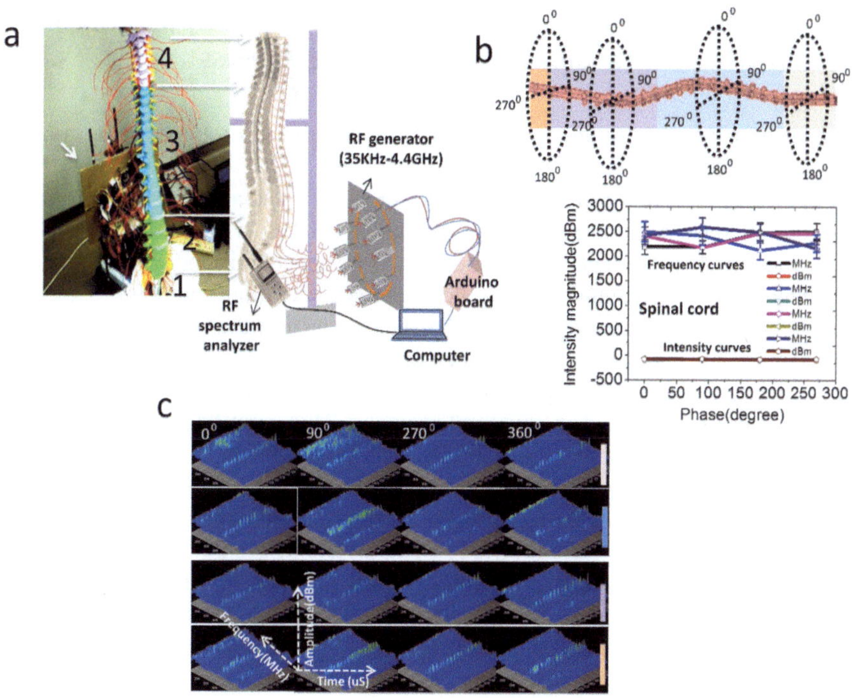

Fig. 6.7 Schematic and background image of the real experimental setup of the spinal cord are shown in panel (**a**), respectively. Thirty-one pairs of handmade spinal nerves are triggered by a channel that includes 10 RF sources (35 kHz–4.4 GHz), and output is detected by an RF spectrum analyzer (300 kHz–4.4 GHz). We observed the resonance spectrums of the spinal cord (panel b: bottom) for 4 various regions (color code—white1, blue2, violet3, and pink4) of at their 0°, 90°, 180°, and 270° angles (panel **b**: top). The intensity and frequency variation and the phase angles for all four colored spinal regions (white, blue, violet, and pink) are shown in panel (**c**)

Electromagnetic resonance is based on standing wave generation. The artificial spinal cord shows an identical function to the real one. Bio-materials geometry is key for a unique pattern of energy distribution over its surface, which is understandable of signal transmission from the spinal cord to the brain or vice versa. We built spinal cord geometry using a structural data file and edited it to make it felicitous to run computer simulation programming to analyze spinal cord functions. We have observed some interesting results related to electric and magnetic field distribution. Spinal cord simulation is performed in the Hz-MHz frequency range using multiple probes at both ends. Resonance peaks are found in the MHz frequency range are the same as those presented in [43]. We got the magnetic field's predominated nature by connecting the probe at the spinal cord's lumbar region.

6.9 Interaction Between the Hippocampus and Spinal Cord During the Instant Decision-Making Process

Instantaneous decision-making in human brain components requires close communication between the spinal cord and hippocampus. Both have one unique geometric similarity; both have a pair of parallel lines of spiral pathways. For the spinal cord, 31 pairs of sensory input and 31 pairs of motor nerves are coupled, while for the hippocampus, two spiral pathways, one input to the upper brain and other output from the upper brain. We compared the biologically accurate hippocampus structure and its handmade dielectric resonator model to understand how the brain circuit for audio visual motion in the humanoid bot brain is affected by mutual interaction between the spiral cord and the hippocampus. Both the brain components regulate their resonant node and anti-nodes based on structural geometry. This is remarkable to observe that length, pitch diameter, and surface density of components on cylindrical area alone regulate the resonance properties of both components.

Throughout the brain, neurons show diversity in their firing frequencies and firing patterns, spike to current injection output, and action potential generation to repeat firing characterized by different latencies [50–52]. Minimum electro diffusion model could explain membrane potential dynamic for showing different firing patterns of [53, 54]. Hippocampus granule cell and spinal motor neurons showed the link between firing pattern and relative ion channel. The spinal cord structure and surrounding elements obstruct spinal cord research, so non-invasive methods are essential. In other words, it is another form of anatomical or geometrical arrangement that is the most challenging task to monitor spinal cords. The function of the spinal cord depends on dorsal root activity that carries the information in impulse form from the brain to body and from body to brain. Inside the head, cranial nerves exhibit function instead of a spinal nerve. The spinal cord is the composition of nerve bundles which are the projection of nerve cells extended from the brain to the lower part of the spinal cord. Information from bodily sensory organs and attached artificial sensors are collected via the spinal nerve and propagated all over the brain. Spinal nerves also send motor information controlled through the cerebellum to the body [29].

Spinal cord inaccessibility produces obstruct in the detection of spinal cord function and its injury effect. For example, to monitor the pain, quality of life, and residual function during spinal cord disease, we need some sensitive methods to represent a structural and neurological function. Imaging the spinal cord is challenging due to its small dimension, inhomogeneous magnetic profile in the MRI system. MRI (magnetic resonance imaging), CT (computed tomography), and PET (positron emission tomography) [55] methods are used to monitor the geometry and function of spinal cords. Currently, we need to improve these methods in terms of high-resolution imaging and time bandwidth. However, progress is being made, but limited groups worldwide are working on spinal cord imaging, but these techniques need to advance for clinical uses. As these techniques advance, our understanding of spinal cord function could enhance and provide a better way to treat disease or injury.

The output of spinal cords effects by its injury. Fracture or disarrangement of vertebrae surrounding the damaged soft tissue and destructions of the axonal signal may be possible reasons for injury. Recently, no pharmacological agents have been available to prevent the growth of spinal cord injury and to regenerate and restore the neurological function of the spinal cord. To activate the spinal cord electrically is the application of noninvasive EMF [56]. The electromagnetic effects on the biological system have been reported long back in medical therapy, leading to health risks. Understanding the electromagnetic field interaction in the human or animal body is key in engineering or medical science. Researchers are currently using the finite difference time domain method to identify the I–V characteristic of the distinct body part. They considered the spinal cord a linear conducting surface. Such a model is capable of generating a significant magnitude of voltage by incident EMF. The spinal cord acts as the frequency-dependent natural antenna. Its resonance response is detected in electric potential and current distribution at 100 MHz by scanning the multi-frequency range (100–2400 MHz) [43]. Radiation emitted from different sources affects the hippocampus dentate and Ammon's gyrus [57]. Similarly, several studies show the detail about the effect of EMF on spinal nerves. Hippocampus memory may be affected by the negative effect of EMF. EMF could alter the geometry and function of the cells [58, 59].

6.10 Communication of Sensory Signals from Skin Cells to Nerve Cells and Brain Upper Region

Old faith regarding the sense of touch was considered as only nerve cells are responsible for sending the information to the brain, but our skin cells, like keratinocytes, equally contribute [60]. Our skin has a complex shape organized by 20 different touch receptors that respond fast to any stimulation. The outer layer (Epidermis) of skin contains fat cells, middle layers (Dermis) have blood vessels, glands, and nerve ending, while the lower layer (Hypodermis) contains fatty tissue.

6.10.1 Various Types of Well-Known Photoreceptors Involve in the Skin and Perform Their Function

Sebaceous gland: It produces the sebum to protect skin and hair. *Sweat gland:* it generates sweat that reaches the skin surface by sweat duct. *Markel's disk*: It is found in the lower and upper dermis. *Meissner's corpuscle*: it is a nerve ending that is found in the upper dermis. *Free nerve ending*: it is a most sensible part of the skin and could sense the light touch, and it is found in the edge of the epidermis. *Adipose tissue*: It is fat storage that supplies the energy to our body. *Bolbous corpuscle*: It looks like capsule, located in the mucous membrane. *Ruffini corpuscle*: It found in middle and lower layers of the dermis.

6.10.2 Sensation Pathways

When photoreceptors of skin cells activate, then sensory information passes through spinal nerve roots and gets integrated by spinal dorsal nuclei or spinothalamic tract. After that, it transmits to the brain region Fig. 6.6a, rightmost), where information passes from the brain stem to the thalamus. Processing is carried out in the thalamus. Then it travels to the cerebral somatosensory cortex [29].

To know which region of the skin network (Fig. 6.8a) is most effective, we alternative put energy sources, P1-midbrain region, P2-left leg, P3-left hand, P4-right leg, and P5-right hand (see Fig. 6.8c). From Fig. 6.8d, the midbrain region effectively transmits energy along the thoracic nerve, hands, and legs. The midbrain is the energy processing unit in the biological brain from which energy goes throughout the body. Other regions are not suitable (see P2, P3, P4, and P5 panels of Fig. 6.8d) for stimulation. The reflection curves for all five ports are given in Fig. 6.8e. We checked energy distributions at all resonance peaks, but we got an effective energy profile at 40.85 MHz.

Skin is one of the most complex organs in the human body; mimicking the skin is challenging. We tried to mimic the human body with midbrain wiring. Skin is mimicked using all types of photoreceptors. Balls, cylinders, spheres, etc. are used to mimic photoreceptor cells in a small cube (see Fig. 6.9a). Cubes cover the whole neural network. It appears in skin form. We scanned frequency spam KHz-MHz and got the effective frequency regions 0–3000 kHz and 0–2000 MHz with multiple resonance peaks. Energy profiles in 2D and 3D forms are detected at those peaks. Again, port 1 is effective (see Fig. 6.9c, d); electromagnetic energy transmits to the body (Fig. 6.9c, d left panel). Energy is less dominated around port 2 (Fig. 6.9c, d mid panel). Some peaks offer a noisy energy profile. When both ports are activated simultaneously, energy is transmitted from brain to body and from body to brain (Fig. 6.9c, d right panel).

6.10.3 How Would We Understand the Interaction Between Spinal Cord and Skin Network by Resonance Frequency Curve?

How does energy synchronous with spinal cord and skin network at different structural scale, we tried to understand. The resonance frequency of skin network and spinal cord are 35 MHz, 40.85 MHz and 2614 MHz, 2310 MHz respectively. Their

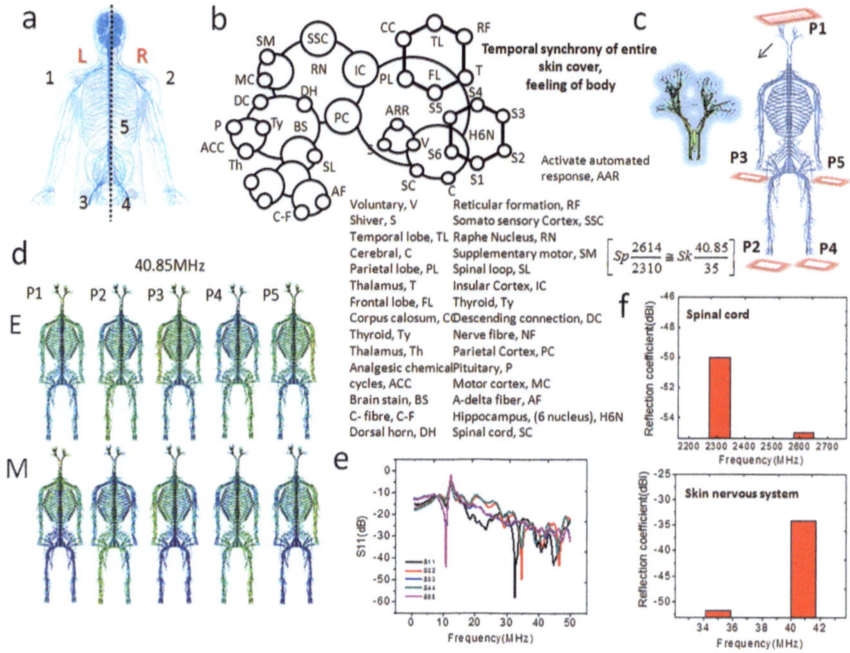

Fig. 6.8 **a** A schematic of the human nervous system shows left and right, LR or C2 symmetry. Furthermore, it consists of 5 distinct nervous regions: 1, 2—left–right arm's nervous system; 3, 4—left–right leg's nervous system; 5—stomach nervous system (panel: **a**). Clocking architectures of temporal synchrony of skin layer and feeling of the body are shown in panel **b**. Dielectric model of the neural network following the complete biological details is created in CST (panel: **c**). Electromagnetic energy is pumped to connectome wires, left—right leg's ends and left—right hand's ends of the human nervous system through ports 1, 2–3, 4–5, respectively. We detected the geometry resonance at 40.85 MHz and obtained electric and magnetic distribution at that resonance peak for all five ports, shown in panels (**e**) and (**d**). Artificial spinal cord has resonance peaks at 2310 and 2614 MHz frequency, and entire nervous system resonance at 35 and 40.85 MHz (panel: **f** top and bottom). Both geometries maintain the same frequency ratio (1.1). **Simulation details**: Used solver—Maxwell equation solver; selected mode—time domain; boundary condition—open space; waveguide port dimension—12 cm × 12 cm; frequency domain 0–50 MHz range

frequencies ratio is almost identical, nearly 1.01 (see Fig. 6.8f). Component geometries are sufficient to perverse information; however, their size at various scale does not means. To be same frequencies ratio, may be one cause to transmit information between two brain organs.

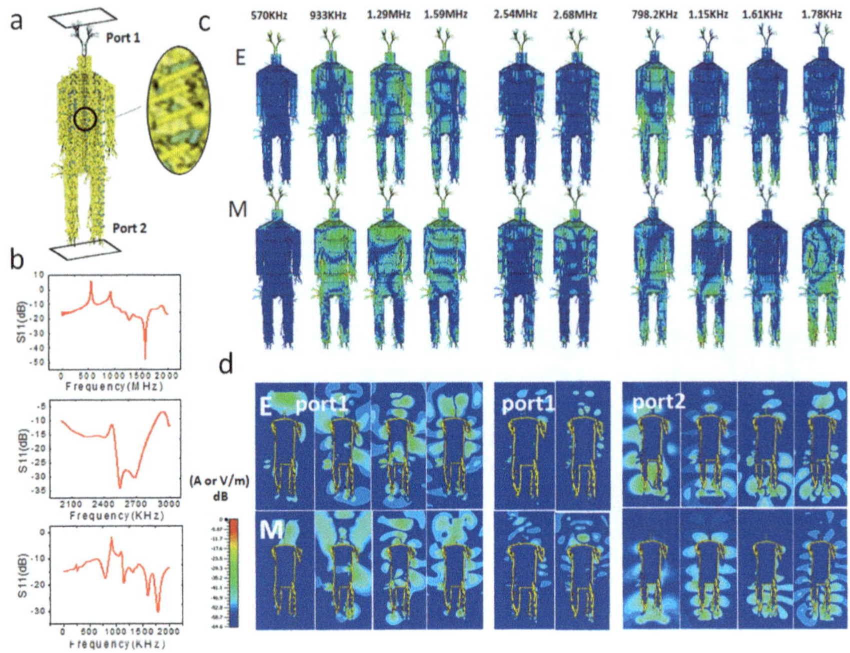

Fig. 6.9 A complex and massive dielectric model of human skin following all packed sensory nerve fibers is created. Mimicked skin network is excited by pumping EM energy through the waveguide ports applied at the bottom and top side of the network (panel: **a**). The simulated resonance spectrum for the individual port; port1 (570 kHz, 933 kHz, 1.29 MHz, and 1.59 MHz), port2(2.45 MHz, 2.68 MHz), and combined ports; port1,2(789.2 kHz, 1.15 kHz, and 1.78 kHz), are shown in panel (**b**). The 3D and 2D EM field distribution of the skin nerve network are shown in panels (**c**) and panel (**d**), respectively. Color code is shown at the bottom left of the panel (**d**). Here, E and M are electric and magnetic fields, respectively. **Simulation details**: Used solver—Maxwell equation solver; selected mode—time domain; boundary condition—open space; waveguide port dimension—16 cm × 16 cm; frequency domain—kHz to MHz frequency range

6.10.4 Could Wireless Communication Affecting the Skin Disrupt Neural Sensory Signals to the Spinal Cord?

As part of a human like bot construction, neural network of a human was built using similar dielectric materials. Skin, the largest organ was artificially created by intricately following the human neural fibers all over the body connected to the touch sensor. The skin-nerve network circuit was cross-checked in the humanoid bot so that it generates a human like input to the spinal cord of the bot. The spinal cord was also created by following the biological details sincerely. Touch, thermal, pressure sensors of the skin of humanoid bot were pumped with 5–20 GHz signals and a noise was carefully generated in the neural network that looked like biological electromagnetic effect. We also created the five sensory signal control circuit, touch and pain signal processing circuits in the humanoid bot brain, like that one in the biological brain and estimated the effect of GHz signal exposure to the decision-making. Humanoid bot is a primitive yet the most meticulously built human signal processing test equipment available to date.

6.11 Cranial Nerve, Spinal Cord, and Human Nerve Network

Sensory pathways in the human body get distracted by a high-frequency signal. We got the energy transmission along the spinal cord, opposite to old belief. According to an old concept, energy flows from top to bottom of the spinal cord, but we found the opposite flow of energy along with it. Variation in energy pattern on bot-brain by triggering sensor at the bottom of the spinal cord confirms that our results are not an artifact. The spinal cord acts as a lossless transmission line. Clocking assembly of skin network and spinal cord offers unknown output by activation and deactivation of clocks. Desired output could be obtained by creating an artificial temporal circuit of spinal cord, cranial nerve, and human nerve network connecting the skin cells. The basic understanding of the spinal cord and skin network may have useful applications in the medical domain.

6.11.1 How to Construct a Clocking Model of Functional Responses of These Structures?

Cranial nerves carry sensory information like touch, balance, image, taste, and sound to our brain. The functional circuit of a cranial nerve is the fusion of image, sound, touch, balance, and smell detector. A clocking circuit means the complication of all signal pathways in the form of a loop. The circle diameter represents the period of the signal, passing through components. Clocks assembly model of pain pathways depicted in Fig. 6.6b. Figure 6.8b shows the temporal map of 'skin network' and 'feeling of the body'(a kind of proprioception), which combine all fundamental elements of the skin network.

Using the literature of cranial nerve, spinal cord, and whole skin geometry and available its 3D model on free accessibly site, we have built 3D model of entire nervous system and skin network where all nerve fibers are in an actual position like literature map. Initially, the skin model includes only the skin's outer surface, but here we mimicked its internal geometry where every fold forms a cavity. We could know the importance of geometry by comparing the results detected between filled cavities and hollow cavities. Triplet of triplet form of cavity exists inside every single biological structure [25, 61]. Energy randomly flows through skin nerve fibers. High and low energy intensity regions are observed there. Theoretical outcomes are verified experimentally using sensors and antenna in humanoid bot built by us.

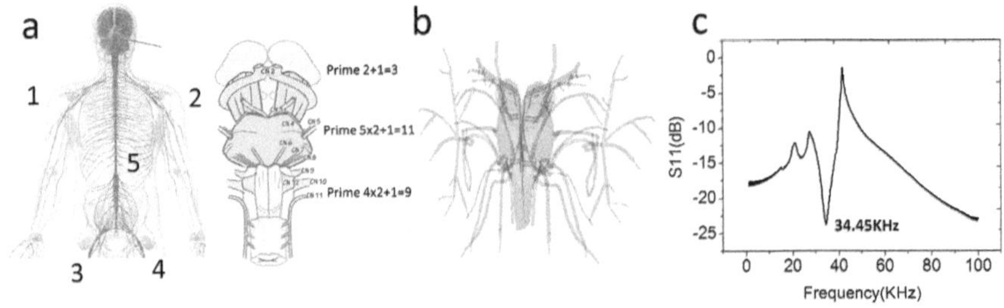

Fig. 6.10 **a** The human nervous system includes five major geometrical regions; 1-left-hand nervous system, 2-right-hand nervous system, 3-left leg nervous system, 4-right leg nervous system, and 5-stomach region nervous system or C5 symmetry. Cranial nerves system; olfactory nerve (limbic area)-I sensory, olfactory (eyes area)-II sensory, oculomotor nerve, trochlear and abducens nerves-V motor nerve, vestibulocochlear nerve-VI sensory, glossopharyngeal and hypoglossal nerves-VII and VIII motor nerve, vagus nerve-IX autonomic, sensory and motor nerve (mixed), spinal accessory nerve-X mixed, facial nerve-XI mixed, trigeminal nerve–XII sensory). It has the left and right or C2 symmetry. **b** A mimicked map of human cranial nerve systems with all 12 sensory or motor nerves. **c** Simulated resonance spectrum of cranial nerves with 34.45 kHz resonance frequency

A schematic model of 12 cranial nerves is shown in Fig. 6.10 b., which has three major cavities based on prime numbers. Replicated geometry is 3–4 times larger than the original. Sub-cavities exist in each cavity. Each sub-cavity built 3 (prime 2 + 1), 11 (prime $5 \times 1 + 1$), $9(4 \times 2 + 1)$ sub intensity regions. Cranial nerves carry sensory information from head to brain. C2 symmetry exists in the cranial nerve, i.e., it appears identical from both sides. EM energy is pumped from its bottom, and the artificial organ shows resonance at 34.45 kHz (Fig. 6.10c). Triplet of triplet form of cavities organized energy distribution in a particular manner. We observed 23 sub-cavities overall cranial nerve. Figure 6.10a shows five symmetrical regions or C5 symmetry in the human nervous system. Initially, all sensory information integrates through 3 cavities and further integrates by 23 sub-cavities of the human head and finally arrives at the brain region. Twelve cranial pathways functionally go out from the head for motor information propagation. Although there are no ions, enzymes, or chemicals, cavities are sufficient to integrate the information.

We have assigned biomaterial as a dielectric resonator. During simulation of biomaterials, its sub-components make unwanted link to each other. In that case, entire geometry acts as single entity. Due to lack of resolution, we could not delete those links. To find out true structural resonance, we have to separate distinct nerve fibers. Replacing those lumpy components, we have prepared neat handmade biomaterials structures like cranial nerve, spinal cord, and skin nerve net (Figs. 6.6a, 6.9c, and 6.11c) by putting each separated nerve fibers.

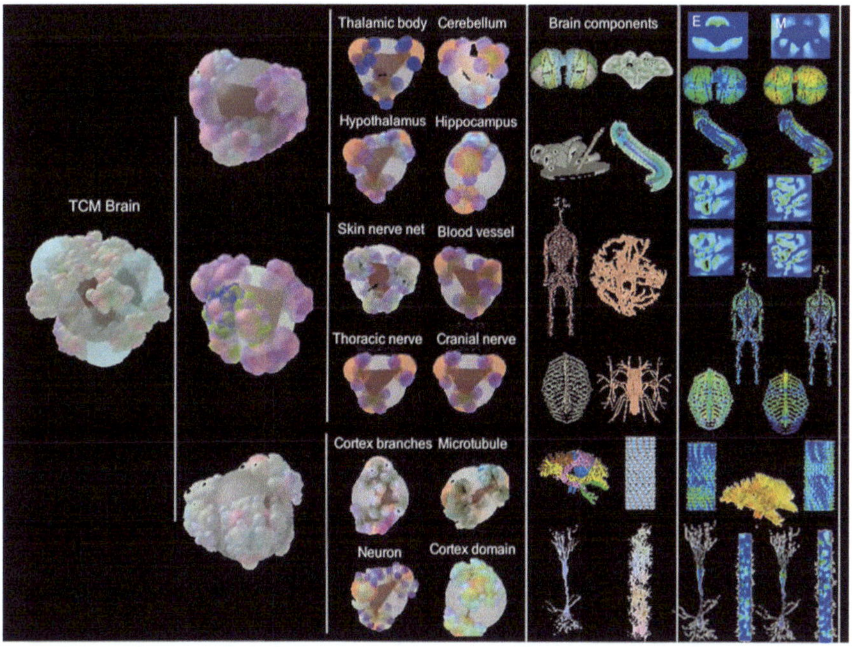

Fig. 6.11 The replicated geometries of all major brain components are shown in column 4, and their respective electromagnetic field distributions appear in column 5. Column 3 presents the 3D clocking assembly of all 12 individual components, while column 2 is about the combined clocking assembly of 4 individual components in 3 distinct lines. When we combine these three units, we get a complete clocking assembly model of the brain

6.11.2 The Role of Geometry and Primes Emulated in Reverse Engineering

Cranial nerve, spinal cord, and neural network make complete nervous system of human body. Geometry of these structures look similar. All these structures are composed of cylindrical tube. Energy distribution looks homogenous on cylinder surface. Symmetry arrangement of cranial nerve (C6), spinal cord (C5), and neural networks (C5) are almost identical.

To find the higher frequency effects on brain region, we attached various sensors (visual, touch, motion, sound, position, and temperature) on bot body. Sensory signal transmits to spinal nerve fibers and make the interference with applied signal in 10–26.5 kHz frequency range (Fig. 6.4). Combine effects observe on bot's brain by EEG machine. We switched on and off sensors (Fig. 6.5) and analyzed activated and deactivated brain region.

When all sensors are off than almost whole region gets silent. Few regions blink due to noise signal. When sensors are on then parietal lobe of bot brain blinks. Touch pathways activates the temporal area while vibration in bot's body or shifts position of bot or varying the surrounding temperature, whole brain area blinks. This artifact occurs by noisy signals which activates extra sensors. These sensors are triggered by human activities resultant touch, motion, position, and temperature sensors simultaneously on. Sound sensors activates parietal and occipital brain lobe.

6.12 The Theoretical and Experimental Challenge in the Creation of the Humanoid Bot Subject, HBS

However, we have built the HBS device for testing our theoretical and experimental results. We have created all individual brain components and tried to learn their functions while simultaneously interactions between brain components are detected by the HBS device. Experimentally realizing the theoretically conceptualized HBS was a challenging task. We wanted only to replicate the 3D clock assembly representing the brain organ.

In theoretical methods, the human nervous system is created by putting tiny-tiny cylinders of different radius. Putting components at the right orientation is a critical challenge in simulation software. In order to create the internal geometry of the spinal cord, we exactly followed biological details [6, 29]. Wrapping a sensor-attached skin around the HBS bot's nervous system has been an effort that took a couple of months.

In the experiment, HBS nervous system is mimicked using a single wire cable. To replicate the same biological geometry in the experiment is an extra challenging task compared to theory. We used a flexible wire which could be bent at different orientations. However, we use paper and glue to hold wire geometry. The spinal cord is a channel of 31 pairs (left and right) nerves. Thirty-one pairs of wires are attached to each sensory nerve. Twenty distinct consciousness circuits are built in the bot's brain so that an operational time crystal that represents all clocks is assembled properly as wired loops. Creating an analogous clock that would replicate the true phase gap has been a challenge. The circuit inside the bot's brain is constructed very carefully; a minor fault in the 3D orientation of the cables may lead to wrong results. We have cross-checked all circuits at the millimeter scale angular deviation while packing the whole neural network and constructing a particular organ. We attached visual, touch, motion, position, sound, and temperature sensors to the body connected with the nervous system. By triggering these sensors by 5 V power supply, signal availability at bot brain is checked by EEG. HBS construction is shown in Fig. 6.4. Several months of intensive hard work were required to prepare the HBS bot.

6.12.1 Detection of Consciousness Features in HBS Brain by EEG

Twenty consciousness circuits are built in the HBS brain, and the Epoc EEG machine detects their output. To check the activation and deactivation of circuits, we have triggered HBS by making the sensors on and off like visual (seeing some object), touch, motion (vibrating the HBS), sound (noise), position (change the position), and temperature (produced the heat near the HBS). All signals are detected by these sensors which are connected to nervous system, signals propagate from sensors to cortex layer of HBS brain through nervous system. Here, cortex layer is mimicked form of PVC plastic. Activation and deactivation effect of these circuits can be seen on cortex layers in terms of brain waves/ neuron firing patterns. Effective change in neuron firing patterns are seen by switching on and off of sensors. From Fig. 6.5, when all attached sensors are off, the neurons fires in an unusual way, a noisy profile appears, when the signal activates, an arranged patterns of neurons appear in respective regions of cortex layers. For example, if we put some colorful objects in front of the HBS eye, the random neuron firing pattern concentrates in visual region, similar thing happens with the touch, motion, sound, and position sensors. In the case of temperature, we have produced the heat near the temperature sensor, in that way, position and visual sensors both get activate resultant combined cortex regions blinks.

A comparative profile of beta waves without noise and with noise is shown in Fig. 6.5. Activate regions of the brain of the bot are noticed by the symbol of letters; L & R—Left & right region of the cerebral hemisphere-parietal and temporal lobe, B& U—back and upper regions of the cerebral hemisphere-occipital and temporal lobe.

Our humanoid bot system is a primitive toy; however, this is a first step towards generating software free testing the feasibility of the brain's information processing model. Here, a few fundamental concerns of futuristic robotics are resolved. First, a cable network, if we built sincerely like a real cognitive circuit, could deliver logical input and output signals for cognitive tasks. Second, the neuronal network of the whole body is very important to build the brain of a futuristic robot; the cortex layer alone is not as dominant. Third, wireless junctions and cables with defined rigidity could partially replicate various brain component responses. Ionic fluids could be avoided. Fourth, the system might learn via itself if the capillary tube's material is replaced by a suitable intelligently designed and synthesized material. Fifth, intelligence is geometric [6, 24, 31, 61], there must have been an unknown geometrical language operation within the real human brain. If resolved, it would help to build a next-generation humanoid bot.

6.13 A Map of Rhythms is Enclosed in Humanoid Bot Brain

Vibration inside the human brain is a special process. It uniquely builds there. At the level of proteins, vibration is variable in the comparative orientation of its substructures. At the level of the neuron, it is the collection of microtubules of different lengths. At the cortical column level, vibration is the structural symmetry of neurons. Therefore, information is a string of rhythms that begin at the smallest oscillator (DNA), end at the largest oscillator (entire brain). The 3D map of the humanoid brain [6, 31] consists of 537 rhythms traced from a standard book of the brain [6, 29] as shown in Fig. 6.11.

Interaction of the electric field with the brain cortex surface is also measured on the humanoid bot's brain. The humanoid bot consists of 20 conscious brain circuits includes the features of all brain components like the hippocampus, cerebellum, midbrain, limbic system, etc., and entire neural body network that includes 31 pairs of the spinal cord, 12 pairs of cranial nerves, and 12 pairs of thoracic nerves. However, we also added five kinds of human sensors; visual, touch, motion, sound, and position sensors on the bot's body. The humanoid bot's natural responses are observed by placing EEG on the bot's brain and connecting a multichannel analyzer (34 channels) to spinal nerves. Planer spiral antennas are positioned on the bot body to absorb the available signal from surrounding space and are also responsible for generating the unified information of spinal cord and brain EEG spectrum. Include all sensors and entire body networks driven by a 5 V power supply from the Arduino board. We recorded the maps of electromagnetic energy distribution on the bot's brain by shifting 16 probes on the surface of the upper cortical layer (thermoplastic) conditionally. All sensors switch and off.

6.14 Conclusion: The Future of Time Crystal Bots

The artificial humanoid bot could be used to study any generic hypothesis regarding information processing in the brain. We have mentioned, this is neither a project to build a lookalike of a human brain nor a true biological organoid. However, it is a geometric replica that attempts all possible vibrational aspects of the brain within a limited frequency scale. While most brain models consider that nerve impulse alone processes the brain's information, we do not fix any mechanism. The resonance of membrane in millisecond's domain is one possibility among various other time domains. Interaction between cortical column assembly and connectome fibers explains learning and decision-making in the brain. Theory and experiment were consistent between two different kinds of hardware. We envision a world of technologies where the time crystal based machines would rule. If ever a commercial humanoid bot is built that runs rhythms similar to 20 conscious experiences by connectome like a conscious living brain, its response would affect a similar cortex.

1. The evolution of virus in Chap. 1 suggested to use self-operating mathematical universe (SOMU), and space–time-topology-prime (STts) metric, to integrate and evolve 3D clock assembly without using any human interference. No software needed, sensors directly connected to the environment converts the signals into 3D clock assembly, which is transported as modulated, phase correlated stream of signals. The signal appears as noise bursts, but a suitable sensor could decode the hidden geometric shapes as the true language for natural programming.
2. The quantum cloaking and anomalous quantum cloaking described in Chap. 2 allow us to build wireless circuits. Of course, we cannot make large scale quantum devices, however, what we could do most is replicating the similar modes of communications. Inside the large humanoid bot subject (HBS), we used dielectric resonators that do not communicate when they vibrate resonantly.
3. Fourth circuit element Hinductor or H-device that we have explained in Chap. 3 was used to build the polyatomic time crystal map of the whole brain body system in Chap. 4. It is the foundation of the 20 consciousness circuits described in Chap. 5, which runs the HBS we have created and described in this chapter. H devices grow within and above and synthesize vortices of different kinds as we have explained in our brain model. In HBS, we have created antenna with a triangular sharp design to generate electromagnetic vortices.

The comparative study between connectome wiring brain and humanoid brain suggests a blind replica of connectome could provide the same EEG output like a combined circuit of 20 conscious rhythms. The similarity in EEG of both kinds of hardware was monitored 24 × 7 for 3 months, we observed changes in the laboratory environment with EEG, multi-channel signal analyzer. Our finding suggests that constructing more advanced versions of a humanoid bot is possible in the future.

References

1. Bandyopadhyay A, Fujita D (2021b) Electromagnetic device, magnetic and electrical vortex synthesis device and magnetic and optical vortex synthesis device; Application no. 2021-172702
2. Bandyopadhyay A, Fujita D (2021a) Method for realizing quantum cloaking in electromagnetic device for remote imaging apparatus; Application no. 2021-172701
3. Bandyopadhyay A, Sahoo P, Fujita D (2021c) Self-learning by information processing device and self-learning for information processing method; Application no. 2021-172703
4. Bandyopadhyay A, Ghosh S, Fujita D (2020b) Universal geometric-musical language for big data processing in an assembly of clocking resonators, JP-2017-150171, 8/2/2017: World patent, WO 2019/026983; US Patent App. 16/635,900
5. Bandyopadhyay A, Ghosh S, Fujita D (2020c) Human brain like intelligent decision-making machine; JP-2017-150173; 8/2/2017; World patent WO 2019/026984; US Patent App. 16/635,892
6. Bandyopadhyay A (2020a) Nanobrain: the making of an artificial brain from a time crystal. Taylor & Francis Inc. Imprint CRC Press Inc., Bosa Roca, United States, p 336. ISBN 10-1439875499. ISBN 13-9781439875490. https://doi.org/10.1201/9780429107771

7. Lancaster MA, Corsini NS, Wolfinger S, Gustafson EH, Phillips AW, Burkard TR, Otani T, Livesey FJ, Knoblich JA (2017) Guided self-organization and cortical plate formation in human brain organoids. Nat Biotechnol 35(7):659–666
8. Lancaster MA, Renner M, Martin CA, Wenzel D, Bicknell LS, Hurles ME, Homfray T, Penninger JM, Jackson AP, Knoblich JA (2013) Cerebral organoids model human brain development and microcephaly. Nature 501(7467):373–379
9. Tan Z, Parisi C, Silvio LD, Dini D, Forte AE (2017) Cryogenic 3D printing of super soft hydrogels. Sci Rep 7:16293
10. Tallinen T, Chung JY, Rousseau F, Girard N, Lefèvre J, Mahadevan L (2016) On the growth and form of cortical convolutions. Nat Phys 12:588–593
11. Albada et al (2018) Performance comparison of the digital neuromorphic hardware SpiNNaker and the neural network simulation software NEST for a full-scale cortical microcircuit model front. Neuroscience
12. Bogdan et al (2018) Structural plasticity on the SpiNNaker many-core neuromorphic system. Front Neurosci
13. Feynman RP (1965) The character of physical law. Cambridge, Mass M.I.T. Press
14. Wheeler JA (1957) On the nature of quantum geometrodynamics. Ann Phys 2:604–614
15. Ascoli GA (1999) Progress and perspectives in computational neuroanatomy. Anat Rec 257:195–207
16. Rall W (1962) Theory physiological properties of dendrites. Ann N Y Acad Sci 96:1071–2000
17. Rieke F, Warland D, de Ruyter van Steveninck RR, Bialek W (1997) Spikes: exploring the neural code. MIT, Cambridge, MA
18. Bi GQ, Poo MM (1998) Synaptic modifications in cultured hippocampal neurons: Dependence on spike timing, synaptic strength, and postsynaptic cell type. J Neurosci 18:10464–10472
19. Markram H, Lubke J, Frotscher M, Sakmann B (1997) Regulation of synaptic efficacy by coincidence of postsynaptic aps and epsps. Science 275:213–215
20. Markram H (2006) The blue brain project. Nat Rev Neurosci 7:153–160
21. Singh P, Ray K, Fujita D, Bandyopadhyay A (2018) Complete dielectric resonator model of human brain from MRI data: a journey from connectome neural branching to single protein. *Lecture* Notes Electr Eng 717–733
22. Striegel DA, Hurdal MK (2009) Chemically based mathematical model for development of cerebral cortical folding patterns. PLoS Comput Biol. https://doi.org/10.1371/journal.pcbi.1000524
23. Terekhovich VE (2012) Probabilistic and geometric languages in the context of the principle of least action. Philos Sci (Novosibirsk) 52(2):108–120
24. Reddy S et al (2018) A brain-like computer made of time crystal: could a metric of prime alone replace a user and alleviate programming forever?. In: Ray K, Pant M, Bandyopadhyay A (eds) Soft computing applications. Studies in computational intelligence, vol 761. Springer, Singapore. https://doi.org/10.1007/978-981-10-8049-4_1
25. Singh P et al (2020) A self-operating time crystal model of the human brain: can we replace entire brain hardware with a 3D fractal architecture of clocks alone? Information 11(5):238
26. Grigorkin AA, Dunaevskii SM (2007) Electronic spectrum and ballistic transport in a helical nanotube. Phys Solid State 49:585
27. Brodmann K (1909) Vergleichende Lokalisationslehre der Grosshirnrinde (in German). Johann Ambrosius Barth, Leipzig
28. Garey LJ (2006) Brodmann's localisation in the cerebral cortex. Springer, New York. ISBN 978-0387-26917-7
29. Carter R (2014) The human brain book: an illustrated guide to its structure, function, and disorders. DK; Expanded, Illustrated, Updated edition
30. Singh et al (2020b) A space-time-topology-prime, stTS metric for a self-operating mathematical universe uses dodecanion geometric algebra of 2–20 D complex vectors. LNNS Springer (in press)
31. Singh et al (2020c) Quaternion, octonion to dodecanion manifold: stereographic projections from infinity lead to a self-operating mathematical universe. AISC Springer (in press)

32. Schacter DL (2012) Psychology Sec. 3:20
33. lversen SD (1984) Recent advances in the anatomy and chemistry of the limbic system. Psychophannacol Limbic Syst 1–16
34. Sherman SM et al (2017) Functioning of circuits connecting thalamus and cortex. Compr Physiol 7(2):713–739
35. McFadyen J et al (2019b) An afferent white matter pathway from the pulvinar to the amygdala facilitates fear recognition. Elife 8:e40766. https://doi.org/10.7554/eLife.40766
36. McFadyen J (2019a) Investigating the subcortical route to the amygdala across species and in disordered fear responses. J Exp Neurosci 13
37. Pessoa L, Adolphs R (2010) Emotion processing and the amygdala: from a 'low road' to 'many roads' of evaluating biological significance. Nat Rev Neurosci 11(11):773–783
38. Gallagher M, Holland PC (1994) The amygdala complex: multiple roles in associative learning and attention. Proc Natl Acad Sci U S A 91(25):11771–11776
39. Gallagher M, Chiba AA (1996) The amygdala and emotion. Curr Opin Neurobiol 6(2):221–227
40. LeDoux J 2(003) The emotional brain, fear and the amygdala. Cell Mole Neurobiol 23(4/5)
41. Pop MG et al (2018) Chapter 1: anatomy and Function on the hypothalamus
42. Arrigo et al (2018) Amygdalar and hippocampal connections with brainstem and spinal cord: a diffusion MRI study in human brain. Neuroscience 343:346–354. https://doi.org/10.1016/j.neuroscience.2016.12.016
43. Balaguru S, Uppal R, Vaid RP, Kumar BP (2012) Investigation of the spinal cord as a natural receptor antenna for incident electromagnetic waves and possible impact on the central nervous system. Electromagn Biol Med 31(2):101–111
44. Thériault G (1992) Electromagnetic fields and cancer risks. Rev Epidemiol Sante Publique 40(Suppl 1):S55–S62
45. Sato T, Adachi Y, Tomori M, Ishii S, Kawabata S, Sekihara K (2009) Functional imaging of spinal cord electrical activity from its evoked magnetic field. IEEE Trans Biomed Eng 56(10):2452–2460
46. Fernandes SR, Salvador R, Wenger C, de Carvalho M, Miranda PC (2017) P084 electric field distribution in the lumbar spinal cord during trans-spinal magnetic stimulation 128(3):48–e50
47. Hock A, Henning A, Boesiger P (2013) 1H-MR spectroscopy in the human spinal cord. AJNR Am J Neuroradiol 34:1682–1689
48. Oliver Wyss P, Hock A, Kollias S (2017) The application of human spinal cord magnetic resonance spectroscopy to clinical studies: a review. Semin Ultrasound CT MRI 38(2):153–162
49. Darabant L, Cretu M, Darabant A (2013) Magnetic stimulation of the spinal cord: experimental results and simulations. IEEE Tran 49(5):1845–1848
50. Kandel ER, Schwartz JH, Jessell TH (2000) Principles of neural science. McGraw-Hill, New York
51. Llinas RR (1988) The intrinsic electrophysiological properties of mammalian neurons: insights into central nervous system function. Science 242(4886):1654–1664. https://doi.org/10.1126/science.3059497
52. Steriade M (2001) The Intact and sliced brain. MIT Press
53. Endresen LP, Hall K, Hoye JS, Myrheim J (2000) A theory for the membrane potential of living cells. Euro J Biophys 29:90–103. https://doi.org/10.1007/s002490050254
54. Herrera-Valdez MA, Smith A, Cruz-Aponte M, McKiernan EC (2011) Biophysical modeling of excitability and membrane integration at the single cell and network levels. BMC Neurosci 12(Suppl. 1):P218
55. Stroman et al (2014) The current state-of-the-art of spinal cord imaging: methods: Neuroimage 84:1070–1081
56. Ross et al (2017) The regenerative effects of electromagnetic field on spinal cord injury. Electromagn Biol Med 36(1):74–87
57. Elfide Gizem K, Berrin Zuhal A, Işınsu A, Kıymet Kübra Y, Adem K, Mehmet Emin Ö (2017) Effects of 900 MHz radiation on the hippocampus and cerebellum of adult rats and attenuation of such effects by folic acid and *Boswellia sacra*. J Microsc Ultrastruct 5(4):216–224

58. Chidiebere EO (2020) Effects of radiofrequency electromagnetic field exposure on neurophysiology. Rev Artic 10(1):6–10
59. Teimori et al (2016) The effects of 30 mT electromagnetic fields on hippocampus cells of rats. Surg Neurol Int 7:70
60. Moehring F et al (2018) Uncovering the cells and circuits of touch in normal and pathological settings. Neuron 100(2):349–360. https://doi.org/10.1016/j.neuron.2018.10.019
61. Saxena K, Singh P, Sahoo P, Sahu S, Ghosh S, Ray K, Fujita D, Bandyopadhyay A (2020) Fractal, scale free electromagnetic resonance of a single brain extracted microtubule nanowire, a single tubulin protein and a single neuron. Fractal Fract 4(2):11. https://doi.org/10.3390/fractalfract4020011

Correction to: The Making of a Humanoid Bot Using Electromagnetic Antenna and Sensors

Correction to:
Chapter 5 in: P. Singh et al., *Biological Antenna to the Humanoid Bot*, Studies in Rhythm Engineering, https://doi.org/10.1007/978-981-16-9677-0_5

In the original version of the book, the following correction has been updated:

In Chap. 5, Reference 29 was corrected as follows:

29. Friston K (2010) The free energy principle: a unified brain theory? Nat Rev Neurosci 11. https://doi.org/10.1038/nrn2787

The book and the chapter have been updated.

The updated version of this chapter can be found at
https://doi.org/10.1007/978-981-16-9677-0_5

P. Singh et al., *Biological Antenna to the Humanoid Bot*, Studies in Rhythm Engineering,
https://doi.org/10.1007/978-981-16-9677-0_7

MIX
Papier aus verantwortungsvollen Quellen
Paper from responsible sources
FSC® C105338

If you have any concerns about our products,
you can contact us on
ProductSafety@springernature.com

In case Publisher is established outside the EU,
the EU authorized representative is:
Springer Nature Customer Service Center GmbH
Europaplatz 3, 69115 Heidelberg, Germany

Printed by Libri Plureos GmbH
in Hamburg, Germany